Writing Maths & Games

Writing Maths & Games

in the open classroom

Herbert R Kohl

Methuen & Co Ltd

First published 1974 as *Math, Writing and Games in the Open Classroom* by *The New York Review*

First English edition published 1977 by Methuen & Co Ltd
11 New Fetter Lane, London EC4P 4EE

Typeset by Red Lion Setters, Holborn, London WC1
Printed in Great Britain by Cambridge University Press

ISBN (hardbound) 0 416 83930 4
ISBN (paperback) 0 416 83940 1

Contents

By the same author

Reading, How To
Age of Complexity
Language and Education of the Deaf
Teaching the 'Unteachable'
36 Children
The Open Classroom
Anthony Cool as Golden Boy

Dedication

to my Mother and Father

Forward to English edition

English readers might be unfamiliar with the age/school-grade relationships in American schools which are mentioned throughout the text. In the United States children enter a kindergarten at the age of five. The first grade is begun at six.

Elementary schools usually have kindergartens and grades 1 to 6 (i.e. children from five to eleven). Grades 7, 8 and 9 (ages twelve, thirteen and fourteen) are combined to form *junior high schools*, while 10, 11 and 12 (fifteen, sixteen and seventeen year olds) constitute *high schools*. Students are expected to attend school from the ages of five to seventeen, and in almost every state are required to attend some accredited school from the ages of six to sixteen. Upon completion of the 12th grade students are awarded a high school diploma.

Another problem that might present itself to English readers is the notion of an open classroom that is referred to throughout the book. In the United States the phrase 'open classroom' is used to refer to an informal classroom which is organized into learning centres and in which the teacher plays the role of facilitator of learning and manager of the centres. Open classrooms have variously been called informal classrooms, child-centred learning environments, alternative classrooms, classrooms with options, and even in some cases free learning situations. There is a great variety amongst classrooms that have been labelled 'open'. Some are highly structured and controlled. These are usually organized into commercially manufactured learning centres which the students have to use at certain times and in certain prespecified ways. Others are much looser and teacher/student created. These classrooms have centres which present materials and opportunities for exploration to students who are free to use them in their own ways. I lean towards the latter type of classroom in my own teaching though most of the material presented in this book can fit in most 'open' situations and a good deal of it may be of use in traditional classrooms.

Herbert R. Kohl
1974

Introduction

Seven years ago I wrote an article for *The New York Review of Books* entitled 'Teaching and Unteachable'. It was the first time I had tried to put in writing some of what I had learned teaching for three years in predominantly black and Puerto Rican elementary schools in New York. When I began teaching I tended to imitate public school teachers I had had and I used their traditional methods. But gradually I moved towards a more informal and open way of working in the classroom. I learned about my students' lives and began to understand how the knowledge and experience they brought to school could be the basis for a curriculum that would, without coercion, interest them in learning. I found that school becomes a far more interesting and valuable place for students if teachers listen carefully to them and do not try to impose on them a rigid curriculum. In the original article I described some of the specific ways of teaching writing that worked in my classrooms and I presented some examples of the children's work.

Recently the editors of *The New York Review* asked me to revise and expand the article. I hadn't looked at it for many years, during which I had directed and taught in a public 'alternative' secondary school and had worked with kindergarten and first-grade students in my daughters' school. I found that my ideas hadn't really changed, but I had been able to test them with hundreds of children in many new situations and had become aware of different approaches and problems that needed fresh discussion.

Creating a new environment for education has become one of my concerns. What do children learn outside school? How can their experiences in school connect with what happens in their lives at home and in the streets? How can the energy, intelligence, and creativity that go into play be brought into the classroom? How can we teach new things to young people without forcing them to perform against their own judgement? What methods will work for teachers in freer educational conditions?

In recasting 'Teaching the Unteachable', which is Part I of

this book, I've tried to emphasize the problem of following up a successful lesson. Like many other teachers I have found that it is fairly easy to produce excitement in the classroom by encouraging children to draw on their own lives and language. But it is much harder to explore a subject in a sustained way so that students can make connections between their own lives and situations that are unfamiliar to them. So I have concentrated here on specific themes for writing that can lead to further possibilities. I've also tried to consider some of the mistakes teachers tend to make and to show some of the ways that writing can be used in other parts of the curriculum.

Part II of the book is about games, play and learning, especially learning mathematics. During the past few years I've found myself using games in every field of study — maths, science, reading, dance, theatre, etc. It became clear to me that young people not only enjoy playing games but like to change them — to create new rules or goals. They like to make up games of their own and teach them to others.

At home, on the street and in school, games are a natural vehicle for learning, but they have not been used explicitly for learning, and so the skills and perceptions games draw on go unrecognized and undeveloped. I have tried to analyse the nature and theory of games in the classroom. I've tried to consider the components of games: boards, pieces, decision-making devices (such as dice and spinners), rules, goals, different concepts of winning. Towards the end of Part II there is a specific discussion of how games can be used in the classroom as well as a suggestion for organizing a game learning centre. Teachers may find it useful to turn first to that section (VI. Using games in the classroom). This section also contains many suggestions and examples that parents should find useful.

Throughout Part II I have tried to show how much arithmetic and mathematics can be learned through games. These of course are not the only subjects that are involved in playing games. However the fact that mathematical skills can be learned in games enables teachers to convince their supervisors that children can learn and enjoy themselves at the same time.

I want to thank those who have given me invaluable

help in writing this book. First there are my children, Tonia, Erica and Joshua, who tried all the games, made up some of them, and taught me a lot about games and gaming. Then there is my wife Judy, who is not much of a game freak but played with us nevertheless. Kathy Sloane, a teacher and friend, went through the manuscript, checked every game, did drawings, edited the manuscript and also typed it.

Vera Mae Fredrickson gave me access to games at the Kroeber Anthropological Museum at the University of California at Berkeley and provided me with invaluable materials on Native American games. Cynthia Brown and Margot Nanny read the manuscript and convinced me that games were indeed important during moments when I wondered myself if play and 'serious' learning could really be done together.

I owe special thanks to the staffs of Pooh's Corner, Honeybear, Birdie, Mr Mopps, and the East Asia Game and Book Center, all in Berkeley. They put up with my frequent wanderings through their toy and game stores and allowed me to open boxes and lay out half-played games on the floor.

Finally I want to thank all the teachers whose work is mentioned in the book (their names are in the text), as well as the staff of the Teachers and Writers Collaborative, and Whitney Ellsworth, Robert Silvers, and Barbara Epstein of *The New York Review*, with whom it is always a pleasure to work.

1 Approaches to Writing

I

SHOP WITH MOM

I love to shop with mom
And talk to the friendly grocer
And help her make the list
Seems to make us closer.
Nellie, aged 11

THE JUNKIES

When they are
in the street
they pass it
along to each
other but when
they see the
police they would
run some would
just stand still
and be beat
so pity ful
that they want
to cry
Mary, aged 11

Nellie's poem received high praise. Her teacher liked the rhyme 'closer' and 'grocer', and thought she said a great deal in four lines. Most of all the teacher was pleased that Nellie expressed such a pleasant and healthy thought. Nellie was pleased too, and her poem was published in the school paper. I was moved and excited by Mary's poem and made the mistake of showing it to the teacher who edited the school newspaper. She was horrified. First of all, she informed me, Mary couldn't possibly know what junkies were, and, moreover, the other children wouldn't be interested in such a poem. There weren't any rhymes or clearly discernible metre. The word 'pitiful' was split up incorrectly, as well as misspelt, 'be beat' wasn't proper English, and, finally, it wasn't really poetry but just the ramblings of a disturbed girl.

My initial reaction was outrage — what did she know about poetry, or about Mary? But it is easy to be cruel about the ignorance that is so characteristic of the schools today. The teacher did believe that she knew what poetry was, and that there was a correct language in which it was expressed. Her attitude towards the correctness of language and the form of poetry was in a way identical to her attitude towards what sentiments good children's poems ought to express. Yet language is not static, nor is it possible *a priori* to establish rules governing what can or cannot be written, any more than it is possible to establish rules governing what can or cannot be felt.

Some years ago, when I was teaching a class of remote, resistant children in a Harlem school, as an experiment I asked the children to write. I expected little. I had been told that the children were from one to three years behind in reading, that they came from 'deprived' and 'disadvantaged' homes, and were ignorant of the language of the schools. I had also been told that their vocabulary was limited, that they couldn't make abstractions, were not introspective, but were accustomed to physical rather than mental activity. Other teachers in the school called the children 'them' and spoke of teaching as a thankless military task. I couldn't accept this idea: I wanted my pupils to tell me about themselves. For reasons that were hardly literary I set out to explore the possibilities of teaching language, literature, and writing in ways that would enable children to express what they felt they were not allowed to acknowledge publicly.

Much to my surprise the children wrote a great deal; and they invented their own language to do so. Only a very small number of the children had what can be called 'talent', and many of them had only a single story to write and rewrite; yet almost all of them responded and seemed to become more alive through their writing.

I have subsequently discovered other teachers who have explored language and literature with their pupils in this way, with results no less dramatic. The children we have taught ranged from the pre-school years to high school, from lower-class ghetto children to upper-class suburban ones. There are few teaching techniques that we share in common, and no single philosophy of education that binds us. If there is anything common to our work it is our concern to listen to what the children have to say and the ability to respond to it as honestly as possible, no matter how painful it may be to our teacherly pride and preconceptions. We have allowed ourselves to learn from our pupils and to expect the unexpected.

Children will not write if they are afraid to talk.

Initially they are suspicious of teachers and reluctant to be honest with them. They have had too many school experiences where the loyalty of the staff and the institutional obligations of teachers have taken precedence over honesty. They have seen too much effort to maintain face and too little respect for justifiable defiance in their school lives. I think children believe that there is a conscious collusion among all of the adults in a school to maintain the impression that the authority is always right, and that life is always pleasant and orderly. Unfortunately, the collusion is unconscious, or at least unspoken. This is dramatically true in slum schools where the pressures of teaching are increased by understaffing and a vague uneasiness about race which is always in the air.

I was assigned to a school in East Harlem in September 1962, and was not sufficiently prepared either for the faculty's polite lies about their successes in the classroom or for the resistance and defiance of the children. My sixth-grade class had thirty-six pupils, all black. For two months I taught in virtual isolation from my pupils. Every attempt I made to develop rapport was coldly rejected. The theme of work scheduled by the school's lesson plan for that semester was 'How We Became Modern America', and the results of my first lesson were characteristic of the dull response everything received.

— Can anyone tell me what was happening around 1850, just before the Civil War? I mean, what do you think you'd see if you walked down Madison Avenue then?

— Cars.

— Do you think there were cars in 1850? That was over a hundred years ago. Think of what you learned last year and try again, do you think there were cars then?

— Yes ... no ... I don't know.

Someone else tried.

— Grass and trees?

The class broke out laughing. I tried to contain my anger and frustration.

— I don't know what you're laughing about, it's the right answer. In those days Harlem was farmland with fields and trees and a few farmhouses. There weren't any roads or houses like the ones outside, or street lights or electricity.

The class was outraged and refused to think. Bright faces took on the dull glaze that is characteristic of the black child who finds it less painful to be thought stupid than to be defiant. There was an uneasy drumming on desk tops. The possibility of there being a time when Harlem didn't exist had never, could never have occurred to the children. Nor did it occur to me that their experience of modern America was not what I had come to teach. After two months, in despair, I asked the kids to write about their block.

WHAT A BLOCK!

My block is the most terrible block I've ever seen. There are at least 25 or 30 narcartic people in my block. The cops come around there and tries to act bad but I bet inside of them they are as scared as can be. They even had in the papers that this block is the worst block, not in Manhattan but in New York City. In the summer they don't do nothing except shooting, shabing, and fighting. They hang all over the stoops and when you say excuse me to them they hear you but they just don't feel like moving. Some times they make me so mad that I feel like slaping them and stuffing and bag of garbage down their throats. Theres only one policeman who can handle these people and we all call him "Sunny". When he comes around in his cop car the people run around the corners, and he wont let anyone sit on the stoops. If you don't believe this story come around some time and you'll find out.

Grace, aged 11

My block is the worse block you ever saw people getting killed or stabbed men and women in buildin's taking dope ...

Mary, aged 11

MY NEIGHBOURHOOD

I live on 117 street, between Madison and 5th avenue. All the bums live around here. But the truth is they don't live here they just hang around the street. All the kids call it "Junkys Paradise".

James, aged 12

My block is a dirty crumbly block!

Clarence, aged 12

The next day I threw out my notes and my lesson plans and talked to the children. What I had been assigned to teach seemed, in any case, an unreal myth about a country that has never existed. I didn't believe the tale of 'progress' the curriculum had prescribed, yet I had been afraid to discard it and had been willing to lie to the children. After all I didn't want to burden them or cause them pain, and I had to teach something. I couldn't 'waste their time'. How scared I must have been when I started teaching in Harlem to accept those hollow rationalizations and use the 'curriculum' to protect me from the children. I accepted the myth that the teacher and the book know all; that complex human questions had 'right' and 'wrong' answers. It was much easier than facing the world the children perceived and attempting to cope with it. I could lean on the teacher's manuals and feel justified in presenting an unambiguously 'good' historical event or short story. It protected my authority as a teacher, which I didn't quite believe in. It was difficult for me — pontificating during the day and knowing at night that I was doing so. How much time could I have 'wasted' even if I let the children dance and play all day while I sought for a new approach? They had already wasted five years in school by the time they arrived in my class.

So we spoke. At first the children were suspicious and ashamed of what they'd written. But as I listened and allowed them to talk they became first bolder and angrier, but then they seemed quieter, relieved. I asked them to write down what they would do to change things, and they responded immediately.

If I could change my block I would stand on Madison Ave and throw nothing but Teargas in it. I would have all the people I liked to get out of the block and then I would become very tall and have big hands and with my big hands I would take all of the narcartic people and pick them up with my hand and throw them in the nearest river and oceans. I would go to some of those old smart alic cops and throw them in the Ocians and Rivers too. I would let the people I like move into the projects so they could tell their friends that they live in a decent block. If I could do this you would never see 117 st again.

Grace, aged 11

If I could change my block I would put all the bums on an Island where they can work there. I would given them lots of food. But I wouldn't let no whiskey be brought to them. After a year I would ship them to new York and make them clean up junk in these back yard and make them maybe make a baseball diamond and put swings basketball courts etc.

Clarence, aged 12

For several weeks after that the children wrote and wrote — what their homes were like, whom they liked, where they came from. I discovered that everything I'd been told about the children's language was irrelevant. Yes, they were hip when they spoke, inarticulate and concrete. But their writing was something else when they felt that no white man was judging their words, threatening their confidence and pride. They faced a blank page and wrote directly and honestly. Recently I have mentioned this to teachers who have accepted the current analyses of 'the language' of the 'disadvantaged'. They asked their children to write and have been as surprised as I was, and shocked by the obvious fact that 'disadvantaged' children will not speak in class because they cannot trust their audience.

Very little that the school offered seemed connected with the children themselves. I read the class novels, stories, poems, brought my library to class and let them know that many people have suffered throughout history and that some were articulate enough to create literature from their lives. They didn't believe me, but they were hungry to know what had been written about and what could be written about.

It was easier for the class to forget their essays then it was for me. They were eager to go beyond their block, to move out into the broader world and into themselves. We talked of families, of brothers and sisters, of uncles, and of Kenny's favourite subject, the Tyranny of Teachers and Moms. We spoke and read about love and madness, families, war, the birth and death of individuals and societies; and then they asked me permission to write themselves. Permission!

In the midst of one of our discussions of fathers Sheila asked me a question that became a symbol for me of my pupils' hunger for concepts. 'Mr Kohl', she said, 'if you wanted to write something about your father that was true is there a word for it?' What she meant was that if there was a word for it she would do it, but if there wasn't she would be scared. One names what is permissible, and denies names to what one fears. Sheila led us to talk about biography and autobiography, and she did get to write of her father.

A BIOGRAPHY OF MY FATHER

My father was born in California.
He wasn't a hero or anything like that to anyone but to me he was. He was a hard working man he wasn't rich but he had enough money to take care of us. He was mean in a way of his own. But I loved him and he loved me. He said to my mother when he die she would feel it. My father was a man who loved his work besides if I was a man who worked at a grocery store I would love it to. He wanted his kids to grow up to be someone and be big at name. He wanted a real life. But when he died we started a real life.

The children spoke of themselves as well. They knew what they felt and sometimes could say it. Sharon came into class angry one day and wrote about a fight.

ONE DAY THERE WAS A BIG FIGHT

One day in school a girl started getting smart with a boy. So the boy said to the girl why don't you come outside? The girl said alright I'll be there. The girl said you just wait. And he said don't wait me back. And so the fight was on. One had a swollen nose the other a black eye. And a teacher stopped the fight. His name was Mr. Mollow. I was saying to myself I wish they would mind their own business. It made me bad. I had wanted to see a fight that day. So I call Mr. Mollow up. I called him all kinds of names. I said you ugly skinney bony man. I was red hot. And when I saw him I rolled my eyes as if I wanted to hit him. All that afternoon I was bad at Mr. Mollow.

I tried to talk to her about her paper, to tell her that 'it made me bad' didn't make any sense. And she explained to me that 'being bad' was a way of acting and that down South a 'bad nigger' was one who was defiant of the white man's demands. She concluded by saying that being bad was good in a way and bad in a way. I asked the class and they agreed. In the midst of the discussion Louis asked one of his characteristically exasperating questions: 'But where do words come from anyway?'

I stumbled over an answer as the uproar became general.

— What use are words anyway?

— Why do people have to talk?

— Why are there good words and bad words?

— Why aren't you supposed to use some words in class?

— Why can't you change words as you like?

I felt that I was being 'put on', and was tempted to pass over the questions glibly; there were no simple answers to the children's questions, and the simplest thing to do when children ask difficult questions is to pretend that they're not serious or they're stupid. But the children were serious.

More and more they asked about language and would not be put off by evasive references to the past, linguistic convention and tradition. Children look away from adults as soon as adults say that things are the way they are because they have always been that way. When a child accepts such an answer it is a good indication that he has given up and decided to be what adults would make him rather than himself.

I decided to explore language with the children, and we talked about mythology together.

I thought of Shelia's question and Louis's question, of Shelia's desire to tell a story and her fear of doing it. The children rescued me. Ronald told me one day that Louis was 'psyching' him and asked me to do something about it. I asked him what he was talking about, what he meant by 'psyching'. He didn't know, and when I asked the class they couldn't quite say either, except that they all knew that Louis was 'psyche', as they put it. I said that Louis couldn't be Psyche since Psyche was female. The kids laughed and asked me what I meant, I countered with the story of Cupid and Psyche, and the next day followed with readings from Apuleius and C. S. Lewis. Then I talked about words that came from Psyche, psychology, psychic, psychosomatic. We even puzzled out the meaning of psyching, and one of the children asked me if there were any words from Cupid. I had never thought of cupidity in that context before, but it made sense.

From Cupid and Psyche we moved to Tantalus, the Sirens and the Odyssey. We talked of Venus and Adonis and spent a week on first Pan and panic, pan-American, then pandemonium and finally on daemonic and demons and devils.

Some of the children wrote myths themselves and created characters named Skyview, Missile and Morass. George used one of the words in his first novel:

> One day, in Ancient Germany, a boy was growing up. His name was Pathos. He was named after this Latin word because he had sensitive feelings.

The class began a romance with words and language that lasted all year. Slowly the children turned to writing, dissatisfied with mere passive learning. They explored their thoughts and played with the many different forms of written expression. I freed the children of the burden of spelling and grammar while they were writing. If a child asked me to comment on the substance of his work I did not talk of the structure. There is no more deadly thing a teacher can do than ignore what a child is trying to express in his writing and comment merely upon the form, neatness and heading.* Yet there is nothing safer if the teacher is afraid to become involved. It is not that I never taught grammar or spelling; it is rather that the teaching of grammar and spelling is not the same as the teaching of writing. Once children care about writing and see it is important to themselves they want to

*The habit of grading a written exercise according to form, neatness, spelling, punctuation and heading is not surprising considering that the written part of the examination for the New York City substitute elementary school teaching licence is graded that way. Content is irrelevant.

write well. At that moment, I found, they easily accept the discipline of learning to write correctly. Vocabulary, spelling and grammar become the means to achieving more precise and sophisticated forms of expression and not merely empty ends in themselves.

In my class a child had permission to write whenever he felt he had to. Barbara, a taciturn girl who had never written a word, put her reader down one day and wrote for fifteen minutes. Then she handed me this:

> ONE COLD AND RAINY NIGHT
>
> It was one cold and rainy night when I was walking through the park and all was in the bed. I saw a owl up in the tree. And all you could see was his eyes. He had big white and black eyes. And it was rainy and it was very very cold that night. And I only had on one thin coat. I was cold that rainy night. I was colder than that owl. I don't know how he could sit up in that tree. It was dark in the park. And only the one who had the light was the owl. He had all the light I needed. It was rainy that stormy night. And I was all by myself. Just walking through the park on my way home. And when I got home I went to bed. And I was thinking about it all that night. And I was saying it was a cold and rainy night and all was in bed.

She explained that she had not slept and wanted to write. She only wrote occasionally after producing this paragraph. I could have encouraged her to continue writing, to build her paragraph into a story. But she didn't want to write. She wanted to exorcise an image that particular day. A teacher does not have to make everything educational, to 'follow up' on all experiences and turn a meaningful moment into a 'learning situation'. There is no need to draw conclusions or summarize what the child said. Often teachers insult their pupils and deceive themselves by commenting and judging where no comment or judgment is necessary.

Larry had been writing voluminous comic book fantasies starring Batman, Robin and the League of Justice. One day he got bored with these heroes and asked me if he could write about himself. I said of course, and so he produced a fragment of an autobiography, after which he returned to his fantasies. He said that the autobiography helped him to invent his own League and his adventurous novels became more personal. I asked him how it helped him, and he said he couldn't say but he knew that it did.

THE STORY OF MY LIFE

Foreword

This story is about a boy named Larry and his life as it is and how it will be. Larry is in the six grade now but this story will tell about his past, present and future. It will tell you how he lived and how he liked it or disliked it. It will tell you how important he was and happy or sad he was in this world it will tell you all his thoughts. It may be pleasant and it may be horrible in place but what ever it is it will be good and exciting but! there will be horrible parts. This story will be made simple and easy but in places hard to understand. This is a nonfiction book.

Where I was Born

In all story they beat around the bush before they tell you the story well I am not this story takes place in the Metropolitan Hospitle.

When I was born I couldn't see at first, but like all families my father was waiting outside after a hour or so I could see shadows. The hospital was very large and their were millions of beds and plenty of people. And their were people in chairs rolling around, people in beds, and people walking around with trays with food or medicine on it. Their was people yelling or praying I was put at a window with other babies so my father could see me their was a big glass and lots of people around me so I could see a lot of black shapes. And since I was a baby I tried to go through the glass but I didn't succeed. All the people kept looking I got scared and cryed soon the nurse came and took all the babies back to their mothers ...

George was shy and quiet and invented his own characters

from the beginning. He was the class artist and drew pictures for everybody. He wrote for himself.

A JOURNEY THROUGH TIME AND SPACE

Chapter III — Just a Tramp

George had been in jail for so long, that he lost everything he had. He didn't even have a cent. "Well" he thought, "I guess I'll have to get a job." He went by a restaurant and got a job as a waiter. One day, a drunky came into the restaurant and ordered some wine. George brought him his wine then after he got through drinking it out of the botle, the drunky said, "How's 'bout yous an' me goin' to a bar t'night?" George was afraid he would lose his job if he had been caught drinking. So he said, "Get out of this restaurant, or I'll call the manager!" With that, the drunken man hit George in the jaw with his fist and knocked him down. George couldn't take being pushed around any longer, so he got up and knocked the drunky down. The drunky got up and pulled out a knife. George grabbed at the knife and tried to make him drop it. They both fought for the knife knocking chairs and taking the worst beatings.

The manager was so afraid that he ran to get the police. Two policemen came in, and the minute George saw them, he knew he would have to spend another month in jail. So he jumped out the restaurant door and ran down the street. The policemen pursued George around the corner where George hid in a hallway and the police passed him. "Whew," he panted quietly, "I'm glad they're gone! But now, I guess, I'm just a tramp. If I leave town, it won't do no good." So he decided to hide in his basement ex-laboratory. He had been in jail for so long he had forgotten where it was. He strolled along the streets day and night. His clothes were getting raggety and people laughed at him. His mother taught him not to beg, even if he didn't have a penny. And George never did beg. And kids made up a song for him:

We know a bum who walks down the street,
In rain, or snow, or slush or sleet
He can't afford to do anything right;
'cause if you see him you'll pop like dynamite!

They made lots more of him like this:

We know a tramp who walks in the damp,
Like a dirty, stinkin' phoney ol 'scamp. —
He can't afford no money at all,
Or have a great big party or ball
'cause he's just a big fat slob,
And never has he gotten a job.

The kids sounded on him every day, and he never did get a decent job. But he still had his mind on being a scientist. To invent things and modernize his country.

There is no limit to the forms of writing that children will experiment with. They will readily become involved in provocative open assignments if they are convinced that the teacher does not simply want a correct answer to an unambiguous question, but rather to hear what they have to say. Themes such as 'On Playing Around' and 'Walls' do not prejudge how a child must respond. 'On How Nice the Summer Is' does. Forms of writing such as the fable or the parable can be open exercises, or predetermined ones. The teacher can provide the framework for many written exercises, but the substance of the children's responses must be drawn from their lives and imaginations.

This is only part of the story, however, the part which can be attempted with a whole class. It is much more difficult to encourage each child to seek his own voice, and to accept the fact that not everyone will have a literary one. It is a mistake to assume that all children have the energy and devotion necessary to write novels or poems. Children select the forms they are most comfortable with, and therefore it is not easy to teach writing. One cannot teach a sixth-grade class to write novels. The best that can be done is to reveal novels to them and be ready to teach those who want to do more than read. I never made 'creative writing' compulsory. Writing must be taught qualitatively — how can one best express oneself, in what way? I found that the children understood these most complex questions and took great pleasure in listening to the various voices of their classmates.

For example, after I read Aesop's fables to my class and we talked about them, they wrote fables of their own.

> Once upon a time there was a pig and a cat. The cat kept saying old dirty pig who want to eat you. And the pig replied when I die I'll be made use of, but when you die you'll just rot. The cat always thought he was better than the pig. When the pig died he was used as food for the people to eat. When the cat died he was buried in old dirt.
> Moral: Live dirty die clean.
>
> *Barbara, aged 11*

> Once a boy was standing on a huge metal flattening machine. The flattener was coming down slowly. Now this boy was a boy who love insects and bugs. The boy could have stopped the machine from coming down but there were two ladiebugs on the button and in order to push the button he would kill the two ladie bugs. The flattener was about a half inch over his head now he made a decision he would have to kill the ladie bugs he quickly pressed the button. The machine stoped he was saved and the ladie bugs were dead.
> Moral: smash or be smashed.
>
> *Kenneth, aged 11*

In writings of this sort we can sense the exhilaration felt by the children in saying things that might have been out of bounds in the atmosphere of the conventional classroom. In fact the conventional classroom itself sometimes becomes the subject of their essays.

> WHY DO RUSSIANS PLANT TREES?
>
> Children sit in classroom waiting for teacher to come. Teacher walks in. Writes on the board. Question is, Why do Russians plant trees?
>
> Some answer are, They plant trees to hide their artillery; To hide their cruelty to the people; So we can't see their secret weapons.
>
> Teacher writes wrong on all. Real answer is Russians plant trees because trees are pretty.
>
> *William Barbour*

WE CAN TEACH EACH OTHER

Inside I feel like I am a nice person; but I have to act like one too. I have to know what kind of person I am. Sometimes, I forget that other people have feelings. Teachers for example, sometimes I hurt them without knowing it. I can say or do something that is so hurtful that they can't say anything but, 'Get up and get out.' They say this so they can go on and teach the rest of us a lesson. But Teachers can hurt you too and they do it just to teach you a lesson.

You are a child of learning in some ways, but in other ways children teach teachers. I don't know what a teacher is like or how she feels, just like a teacher doesn't really know what I'm like or how I feel. So I can teach her what and how I think and feel. Teachers have been children before, but they seem to forget what it's like because the time changes in the way that the weather changes. So they can't say, well it's like when I was a child.

No, it's not like that, because people are changing and our minds are changing too. So children teach teachers a new lesson, about children today.

Patricia Williams

2

There are many ways to introduce writing into the classroom. I have taught small writing seminars in which I presented themes and encouraged people to read their work aloud. I've also taught in more open settings where I acted primarily as a resource, to answer questions, set themes and step out of the way. I feel most comfortable in a situation where many different kinds of activities take place — where students and teachers meet regularly to read their work and suggest new themes and ideas as well as to work independently or in small informal groups.

When I first encounter a new group of youngsters I don't want to assume anything about what they can or can't do.

I want to discover who they are while they discover me. Therefore I start by presenting a simple familiar object or situation that we can look at together in different ways.

For example once with a group of fourth graders I started by giving all the students a blank piece of paper and asking them to imagine that it was a wall. Then I suggested they write on it. There were giggles, some embarrassed looks — one student asked me if she had to write anything and I said it was up to her. Slowly the students began to write.

I wrote too, and after a while we shared our walls, reading what we had written and talking of the kinds of things we had thought of writing but hadn't dared to put down. Then I put a large piece of paper on the floor and suggested that we think of it as a public wall, that it belonged to all of us and was seen by all of us. Then I suggested that we write on it together. The children liked the idea and wrote names, nicknames, jokes, slogans. Some drew pictures and others wrote over the drawings or changed names or added to nicknames. My own name was written:

Mr. Kohl love vanessa

and changed:

Mr. Kohl is Mr. Cold

and changed again:

Mr. Cold is Mr. Cool

and played with:

Mr. Cool Cold
thinks hes a teacher

There was so much on the public wall that it took a while for us to read it. We didn't talk about what was written, I didn't make a lesson out of it, but a great deal of writing, revising and concentration on words and meanings was taking place.

That was the entire first session, which gave me a chance to see what the children could do, while they had a chance to feel me out. There was no formal testing nor was there any compulsion involved. On the other hand the subject seemed sufficiently interesting and suggestive that the children participated willingly.

In an open classroom the introduction of a new subject or activity has to be thought through carefully. The more school learning and facts and skills the introduction presupposes, the more the students are likely to feel tested and to become anxious and unenthusiastic. In introducing a writing programme a teacher ought to begin by discovering what the students' lives are like, encouraging them to make use of their experience at home, with their friends, in the street.

A blank piece of paper can be used in many ways. The way I began does not have to be the way in which other teachers begin their writing programmes. The poet Muriel Rukeyser gives her writing classes a blank piece of paper and asks them to do something with it — anything. It is impossible to predict what the students will do. They make paper aeroplanes, rip the paper to pieces, or even write poems. Whatever happens they talk about it and look at the variety of responses that were produced. From there they go on to more elaborate writing themes such as:

> A story I could not tell

or

> A secret letter.

There are other ways to begin writing programmes. One can start with the students writing on such themes as 'my block' or 'tricking and being tricked' or 'playing and playing around'. One can also start with the lyrics of pop music or of the blues or of rock or soul music; with street signs, parodies of TV ads, lists of names and nicknames, etc.

I like to tell stories and often begin with some silly or funny or moving stories of my own and encourage the students to swap stories and jokes and tales of their own.

One of my favourite stories is about the wise men of Chelm, a mythical Yiddish town where everyone is a fool:

A ROAD AND A HOSPITAL

In Chelm, there was a road, a very good road, that led way up, high up, into a mountain and then it came to a sudden stop and beyond was a steep precipice. When people were careless or not watchful, they would fall and they were hurt some very badly.

Mothers warned their children to be careful, but when they became engrossed in play, they would forget, and they, too, would fall.

'This is a curse', said the people of Chelm.

They went to their Wise Men.

'You must do something', the mothers cried. 'Not a single child is safe.'

The Wise Men listened, stroked their long, white beards, and said: 'Yes, Yes. Quite so.'

'Let our Wise Men think', said the people. They tip-toed away so that there would be quiet.

Day after day, week after week, month after month, the Wise Men thought and thought.

One day all the Wise Men appeared before the people of Chelm looking very pleased and contented with themselves.

'Our Wise Men have been working hard. They will know what to do', said the people.

And sure enough they were right.

'At the bottom of the mountain', said the wisest of the Wise Men, 'we will build a hospital and when our good people fall, the doctors will be right there to take care of them.'

And so it was done.

At the bottom of the terrible precipice, the people of Chelm built the biggest and finest hospital in any town, near or far.

When the hospital was finished, bright and glistening in the deep valley, the Wise Men stroked their beards vigorously, smiled contentedly, patted one another on the back in pride.

'Ah,' said the people, 'what other town has Wise Men as wise as ours?'

After telling the story we start talking about fools, about imaginary towns where everyone is a thief or a trickster or a hustler or a liar; about places where people laugh when they should cry and cry when they should laugh. Sometimes students readily volunteer stories of their own, make up and name towns like:

Hustings where everyone hustles

or

The Village of Deceit

or

The City of Crook

and begin to generate tales and characters. Sometimes it is good to tape the students' tales, other times it makes sense to encourage them to write the best ones down. However, there are times when the telling of the tale, the performance, is the key thing and to suggest writing destroys everyone's enthusiasm. There is no formula for knowing when to move from the spoken or taped to the written word, and different possibilities have to be explored. Sometimes the students will write in a frenzy, at other times it will make sense to select a few favourite tales to transcribe. Occasionally the students will want to move in an unexpected direction and a teacher has to learn how to follow the hints and leads the students present and be prepared to abandon even the cleverest and favourite plans and ideas.

I like Yiddish tales. Other teachers have found Sufi teaching tales to be effective.

Brer Rabbit, Anansi, Coyote and other trickster figures seem to interest most young people who are always concerned with tricking people and being tricked, with developing their wits and figuring out how to beat adults at their own games.

Young people also love to hear about the personal past of the adults around them — about what teachers were like as children, about their families, the adventures the adults had and the mistakes they made. I've gone on for hours about my past to kids who keep throwing questions:

What is your mother like?
Did you ever love someone who didn't love you?
Did you ever steal anything?
What was your first job?
Were you poor or rich as a kid?

Young people like to hear and create tales about imaginary families as well as historical ones, about being descended from certain gods and goddesses, about creating imaginary good and evil parents, or pretending they were descended from certain animals or are reincarnations of famous people or heroes. Talking and writing often go on together — the group elaborating on group themes and creating collective stories; individuals going off to make their own stories and myths.

It is possible to initiate writing programmes by reading

other people's work as well as by telling tales or presenting themes. A few years ago I did a seminar in writing with some junior-high-school students. At the first session we sat around a table and talked about various forms of prose, and I read some selections from Hemingway, Faulkner, Thomas Mann and Herman Hessé. I asked the students to listen to the selections in order to discover the voice of the author instead of the plot or the characters. Then I asked the students to imagine that the room we were in was dislocated, that we were spinning in space and that all of a sudden we landed ... and opened the door to the room. I then asked them to write a description of what they saw.

All but two of the students began writing immediately. One girl said she didn't feel like writing, that there were other more important things on her mind and that she couldn't concentrate. She asked if she could just sit and watch, and I told her of course she could. Another student, a boy, sat on the floor (we were all on the floor sitting in a circle by that time) and played nervously with his pencil. There was sweat on his forehead and it was obvious that he was experiencing extreme anxiety. I asked him what the matter was, and he said he just couldn't write although he wanted to. He didn't know how to begin, I suggested he free associate, put down any words that came into his mind. I also explained how difficult it often is for professional writers to get started. A blank piece of paper is a very threatening object. It is an open invitation to create that gives hardly any clues to how specifically one can begin.

The rest of the students finished their papers. Some were several pages long, some only a few sentences. One was a poem. I asked the students to read their papers and explained that there would be no grading. I also explained that learning to write well often involves writing poorly for a while. There is no point in grading or condemning bad writing. There is a point to working on it. Writing well is discovering one's best written voice and is a part of discovering oneself. It takes time and hard work.

No one volunteered to read. Everyone looked away from me. I had the feeling that they felt I intended to humiliate them. I asked again and one girl said she couldn't read her piece because her writing was awful. Some of the others muttered in agreement. Almost without exception the students were all convinced that they could not write well. They were tearing themselves down, frightened of their own voices. The next stage would obviously be that they would hate to write and only do so when forced; it was boring and laborious and it frightened them.

I finally got one student to agree to read. Before she began I asked the class to listen not to the words or the grammar or the ideas but to the voice of the author. We listened.

> A plain. Lonely. A man walks across with a stick in his hand. He looks around. Nothing. Nothing. Long lonely empty.

We listened to her voice, to the pleading, to the condensed adjectiveless speech, and spoke of the style she was struggling to achieve. Then someone else volunteered to read:

> Lushness, a smooth shiny water. Animals all about music in the air, sweet harmonious sounds a flowing rippling undulating ... I can't go on

The voice here was quite different, and the student couldn't finish the piece because she couldn't sustain the tone. We talked about the voice in this short piece and how it seemed artificial, what the teacher might want because it used many big words and expressed a positive happy attitude. The voice the student chose trapped her into being artificial. She was trying to figure out what voice would please someone else rather than discover one of her own. A third essay was more direct than the other two and hit harder:

> The door opened. Shit! It was the same old crummy world, the dirty streets, the people hanging around drinking shooting stabbing. The room should never have stopped spinning free in space, free at last.

The students decided that the voice in this piece seemed like the person who wrote it. We talked and talked about voices and all of the students did come round to reading their papers. They began to listen to each other's papers, not in order to criticize them or place a mark on them or compare them or rate them, but in order to understand them and learn something about each other.

It is absurd that young people fear writing, are ashamed of their own voices. It is important to encourage them to listen to themselves and to each other.

At the second meeting of my class I picked up a random selection of novels and read only the first lines. We tried to figure out the voice of the authors from the way they introduced themselves to the reader. Here are a few examples:

> She stood at the gate, waiting; behind her, the swamp, in front of her, Colored Town, beyond it, all Maxwell. Tall and slim and white in the dusk, the girl stood there, hands on the picket gate.
>
> (*Strange Fruit*, Lillian Smith, 1944)

> Dinner with Florence Green. The old babe is on a kick tonight; I want to go to some other country, she announces. Everyone wonders what this can mean.
>
> (*Come Back, Dr. Caligari*, Donald Barthelme, 1964)

> It was the best of times, it was the worst of times, it was the age of wisdom, it was the age of foolishness, it was the epoch of belief, it was the epoch of incredulity, it was the season of Light, it was the season of Darkness ...
>
> (*A Tale of Two Cities*, Charles Dickens, 1868)

> Now that we are cool, he said, and regret that we hurt each other, I am not sorry that it happened.
>
> (*Green Mansions*, W. H. Hudson, 1904)

> The crowded red double decker bus inched its way through the snarl of traffic in Aldgate. It was almost as if it was reluctant to get rid of the overload of noisy, earthy char-women it had collected on its run through the city ...
>
> (*To Sir, With Love*, H. R. Braithwaite, 1959)

3

It is important for students to have the opportunity to understand writing and telling stories and tales in this fluid way. There have to be places and time to write in private, there has to be lots of paper around and pens and pencils.

There are classrooms where each student is responsible for providing his or her own pen, and everyone is allotted a few pieces of paper a day. Students without pens have to beg them from other kids or sit quietly doing nothing.

I have always found it useful to have boxes of inexpensive pens and pencils available as well as lots of cheap paper of different sizes and shapes. Often I buy some of the material out of my own pocket and encourage the students to look around their homes and bring in as many pens and pencils and crayons as they can; not for each individual to maintain in the secrecy of a cubby or desk, but to add to a class box so everyone will have access to writing implements when they need them.

4

In many open learning situtations there is no quiet and private place for young people to read or write or solve problems. When I was a student in the sixth grade the teacher used to propose special maths or grammar problems and give us a half hour to solve them. Any student who got the right answers in that time was given a special high grade. I never got any of the problems right, and felt I was very stupid. I would take the special problems home with me, hide in my room and think about them. At home, alone, able to listen to music and walk around, I could solve the problems. In school with all the other students around me, with the silence in the room and the pressure of grades and time, I could only worry about being thought stupid by the teacher and the other kids.

When I taught the sixth grade there were many fights in the class. Students often defied me, but more often they fought with each other. I had to resolve the fights as quickly and arbitrarily as possible. There was no place to go to work out the problems behind the fights or let the combatants rest and talk to each other.

I remember a particularly bad fight between two boys who

had liked each other very much up until the moment the battle broke out in the classroom. I wanted to talk with the boys but that was impossible. There was no one to watch my class, and even if there were, the only place wc could talk was in the hall, which was full of movement and not at all private.

There are certain things best done in private, among them thinking, creating, solving problems and conflicts, and resting.

Thinking is a complex activity that no one has been able to describe fully. It is clear however that different people find themselves able to think best in different circumstances. Some people can think best with music on, others need complete silence. Some people do their best thinking while walking, others while sitting in an easy chair, or lying down, or sitting in a restaurant or a public square watching people they don't know pass by.

Problems are solved in many different ways. Some people dream up answers, others need a pen in their hand or another person to push them into thinking of a solution.

Making a work of art is an even more eccentric activity. Artists and writers have special conditions and environments under which they work. Some can work with other people, others need to be completely alone. I cannot write with anyone else around and I have to write on a certain size paper and with a special pen. This is not just my madness. We all seek the most comfortable environments in which to work. Robert Creeley, the poet, has expressed what I am getting at in a speech, given at a poetry conference in Vancouver, about the settings in which poets created their work. He said:

> What I'm trying to say is don't start thinking of writing as some particular activity leading to some particular effect for some particular purpose. It is just as relevant what size paper you use as whether or not you think you are writing a sonnet. In fact, it's more relevant. And this aspect of your activity ought to be, you ought to be aware of it ... In other words if you want to write with a paper like this, please DO! If you find yourself stuck with habits of articulation try doing something else, try shifting the physical context ...
>
> (Robert Creeley, 'Context of Poetry',
> *Audit*, vol. V, no. 1, Spring, 1968)

Solving conflicts is no less personal than creating a work of art. I do not feel easy dealing with someone's problems in public, and as a teacher I was always embarrassed by the need to deal with the problems of two or three students with all the other students present.

It is important to build private spaces into existing classrooms. Right now I am experimenting with the possibility of creating private corners and little rooms within existing rooms. Material is magical. A few muslin dividers hanging from the ceiling can create private spaces. Clothes closets can be turned into private rooms. Even rugs on the floor can mark private spaces. The hall outside a classroom can be partitioned off and turned into a series of private spaces as well. Portable Japanese walls or ingenious room dividers can be used. The classroom can be changed into a complex environment accommodating and respecting private as well as public experience.

5

In kindergarten and first grade, where the students cannot be expected to read or write with fluency, collective storytelling and book making are effective ways of getting young people involved in writing. For example it is possible to sit around in a circle in front of a blackboard and begin by putting a name on the board and telling students that they can made a story by each contributing a word. Here is a story that developed in a first grade in Berkeley. The teacher began with the name John, which evolved into the following short tale:

> John went to the zoo
> Where he belongs.

Another story began with the name Lizzie and ended up:

> Lizzie wants a big
> cadillac and she got
> two and drives around the
> school all day.

After the story has been finished a number of students usually want to read it, even though they might not know how to figure out the words if they saw them in a different context. I encourage the students to read the story over and over, to practise and to take pleasure in playing with their own creation. Often it makes sense for the teacher to put the story onto a photocopier, run off enough copies for the class or even for the school, and let the students illustrate their work and send copies around the school as literary broadsides.

Generating a collective story can be fun for older people as well, and I've used the exact same technique with high school and college students. However, words like dope and wine and deceit and hypocrisy usually come in pretty fast, so that a story beginning with the name John might look like this in high school:

> John bought a kilo of bad
> dope and kicked the ass of
> his mother who sold it to him.

or like this in college:

> John the Apostle laughed to
> hear that Satan visited the
> earth and discovered it was
> worse than hell.

Another technique that works particularly well with young people could be called 'the continuing story of our very own hero and heroine'. It is possible to talk to young children about the powers they would like to have, the adventures they can imagine, and enable a group to develop composite figures who can be turned into drawings or dolls and carried through all kinds of adventures. For example such figures as

> superchild
> submarine girl
> The Rainbow Power Kid
> Braino the evil genius
> Willful — the Wicked Kid

can be carried through continuing adventures with different children taking turns setting the scene or playing villains. These stories can be done through dramatic improvisation;

they can be told and taped or turned into comic strips or even film strips or TV tapes. The greater the range of activities involved in creating, enacting, recording and shaping a tale, the more different students have opportunities to use their particular skills and discover their talents.

My daughters Tonia and Erica, who are six and four-and-a-half, had a superheroine, Melissa May Anne Fingerwald Lewis Pampelmouse, alias Raccoon Girl. For a while they told a different story every night about Racoon Girl, who wasn't afraid of the night, sneaked into people's houses when they weren't home, and climbed trees the way most people walk up hills.

Young children also love to make up alphabet, beast, monster and silly creature books. I have a collection of rhyming alphabets, bestiaries, monster books, comic ABC and number books which I show to young people who immediately get the point and rush to make their own. Last year at Wari school in Berkeley a group of fourth, fifth and sixth graders created the following monsters:

> F: the fire monster
> L: the luxury loony lazy life lobster
> M: the money monster who always carries around 500 hundred-dollar bills

and finally

> O: the oil monster who loves to pollute water.

Some of my favourite alphabet and monster books are:

Baldwin, Ruth M., *One Hundred Nineteenth Century Rhyming Alphabets in English* (Southern Illinois University Press, 1972, $15.00).

Baskin, Leonard, *Hosie's Alphabet* (Viking, 1972, $4.95).

Borges, Jorge Luis, *The Book of Imaginary Beings* (E. P. Dutton, 1969, $7.95; Avon, $1.45, paper).

Cortázar, Julio, *Cronopios and Famas* (Pantheon, 1969, $4.95).

Schmiderer, Dorothy, *The Alphabeast Book* (Holt, Rinehart and Winston, 1971, $3.95; $1.25, paper).

Sorel, Nancy Caldwell, *World People* (American Heritage Press, 1970, $6.95).

The Studio Book of Alphabets (Viking, 1963, $2.00).

6

Sometimes a teacher might want to explain a certain form of writing to a class — a fable or sonnet or parable or sestina. Sticking to the traditional texts makes it seem that there is only one proper or best way to introduce each form. This is nonsense. There are many different ways to approach the sonnet or sestina, for example, and often the word sonnet or sestina might be the last thing one mentioned. Two teachers introduced their students to the fable in the following ways.

Lila Eberman, who was at that time teaching at Benjamin Franklin High School in New York, and Elna Wurtzel, who was teaching sixth grade in a public school in Manhattan, were both members of a seminar on writing I conducted for the Teachers and Writers Collaborative in New York. During the seminar the teachers and writers discussed various literary forms and tried their hands at writing them themselves. Then they reinterpreted what had been discussed and done during the seminar into specific strategies that could be used in their own quite different classrooms. Here are some excerpts from the teachers' diaries. Lila Eberman at Benjamin Franklin High School wrote:

TWO LESSONS: INTRODUCING A UNIT ON THE FABLE

Taught at Benjamin Franklin High School to two tenth-grade classes of children in the College Bound program. There are 17 pupils in each class.

I decided to begin the unit by giving for homework an assignment to copy down everything the students saw written on walls, sidewalks, subway posters on their way home from school that day. Both classes responded by laughing:

'Everything?'
'You don't mind foul words?'
'There's a *lot* of poets on my block!'
'Where should I look when I read them?' (aloud in class the next day)
'Your mind's goin' to conk out.'

One boy remarked that he knew that there was a name for these writings but couldn't remember what it was. I told the class it was 'graffiti' and wrote the word on the board. The boy then said that he thought that there was a man doing a book on graffiti — was this for that book? I told him that it wasn't, but that we were going to use the graffiti for a lesson. The general feeling among the students was that it was an unusual homework assignment, but they were enthusiastic about doing it.

Lesson 1
The aim of this first lesson was to get across the idea of people expressing thoughts and feelings through writing. First there was a recitation of the graffiti they had seen. All were anxious to read what they had copied down — but they did not want to read the most vulgar. We then discussed the reasons that people write on walls, the general conclusion being that the writers wanted to reach a wide audience. We made a frequency chart describing the subject matter of the writing:

Love messages/insults/bragging/the truth about people/sex/
(real and puppy love) (to the reader and others)
humour/politics/foul language/hatred/identification of territory

As to the type of language used — the pupils said that slang or 'hip' language was used because everyone understands it. After the discussion, I told them that they could go over to the blackboards and write whatever they wanted. One class jumped right up, but the other class was slightly shy and didn't begin until I had personally given each student a piece of chalk. They all enjoyed themselves, finally, laughing, talking and reading one another's graffiti. Here are some of the results:

> Your mother plays drums
> With the midnight bums
>
> Your father is a nice lady
> False teeth are what's happening
> Your father works in the marqueta
> Junky go home
> The junkies are coming

Your mother plays
shortstop for
the hunchback dodgers

All of these turned out to be in the category of insults but the humor and imagery are quite imaginative. One pupil remarked that the writings reminded him of badges he'd seen around.

After they sat down, I showed them the covers given out by the Teachers and Writers Collaborative behind which they were to keep their writing. On the front is a photo of a written and drawn-over wall. I suggested that if they wanted to, they could turn the sheet over and make their own wall and use it for the cover. I was then asked whether we could have a paper put upon the bulletin board on which the students could write. I agreed that it was a good idea.

After some thought I decided that this lesson, though it bore little relevance to fable writing, was a good one in which to present the idea that original or creative writing by 'ordinary' people — young people especially — is fun and interesting to do and to read. In this, I feel, is the value of a lesson of this type.

Lesson 2

In this lesson I wanted to introduce fable writing to the class. I put on the board 'All that glitters is not gold' (because it was the first one I thought of) and began by asking the class what kind of statement this was. They said that it was a proverb or saying or moral, which was defined by the pupils as something that tells you about life so that you learn something, so that you don't make the same mistake. It tries to teach you a lesson.

The students gave examples of other proverbs that they knew and we listed them on the board and discussed each one. They came up with the most common — Don't put all your eggs ... Don't count your chickens ... You scratch my back ... etc. One student contributed an original: Do unto others' mothers before others' mothers do unto you. One boy remarked that these sayings were not always true — for instance, regarding 'Counting chickens', he always does chores for his mother and he figures out beforehand how much money he'll earn and how he is going to spend the money — and he is usually right.

One night a boy called Herbert walked through the streets, until all of a sudden the peacefulness of night was broken when he saw a group of about 20 charge at him. He, not knowing what to do, stepped to the side because he thought that maybe they were not after him. But one of the gangsters called out there he is, then Herbert began to run for what seemed to be his life. But to no avail, the gang caught up to him and gave him a beating like never before. The police and the ambulance were called to pick up what was left of him. He was hospitalized for two months. After he came out of the hospital he said he was going to get even, so he bought himself a tonson machine-gun. That night he waited where the gangs hang out. Before he knew it the gang appeared coming up the block. Herbert was waiting and ready. As they passed by he came out shooting madly at each and every one of them. None of them survived. The police caught him and arrested him. Soon he was brought to trial. He lost the case. Before his sentence was to be announced the Judge said do you have anything to say, and Herbert said do unto others as others do unto you.

Nothing in life is to be feared, it is only to be understood.

One day an ostrich named Johnny wondered why he did not fall off the earth and when he went to bed he knew that he was off his feet so he would tie himself to the bedpole at night. One night when Johnny went to sleep he saw a vision of an angel named Leonardo, and the angel told Johnny that there was nothing to fear and that an angel named gravity was holding you down. And from that time on Johnny never feared anything again.

Elna Wurtzel began in a completely different way with her sixth graders:

NOTES ON FABLE LESSON 1

The purpose of this first lesson was to concentrate on stories which end with a moral, or as I called it, a 'message'.

I wanted to emphasize that the messages did not have to conform to 'good thought' or 'good behaviour' usually demanded of children. Therefore I told the following three stories.

Story 1
(in summary). Peter Rabbit goes out to play. His mother says 'don't go in Mr. McGregor's cabbage patch'. He plays around and is finally tempted into the cabbage patch. He enjoys it, eats cabbages, and falls asleep. Mr. McGregor captures him. At the last moment he escapes.

I asked the class what message we could add as a last line on this story. Many hands went up. Everyone seemed pleased to feel that he knew the message.

The Messages:

1) Now listen you guys, don't go outside without your mother and father.

2) When your mother says don't go in the cabbage patch, don't go in the cabbage patch — and don't you forget it.

3) Stay out of cabbage patches.

4) Keep out of other people's property.

5) Don't go where your mother says not to.

Story II
Peter Rabbit — same beginning as above. Once again Peter Rabbit is tempted to go into the cabbage patch. He plays around, eats cabbages, and begins to feel sleepy. Then he says, 'Boy, I'm not going to be a dumb bunny and fall asleep in the cabbage patch!' So he sneaks out again, sleeps somewhere else, and finally goes home. 'What have you been doing?' asks his mother. 'Oh, nothing', he replies.

I asked the class what message we could add on as a last line. The children were somewhat confused, as the usual moral was not so clear to them.

Someone said 'You shouldn't tell lies to you mother'. I asked if that was what the story taught us. She said no.

Several other children wanted to add on to the story so that something bad would happen to Peter Rabbit so that we could then have the ending, 'You shouldn't tell lies to your mother'.

I told them to try to stick to the story just as I had told it. Was there any message in that story? Two children volunteered,

1) Be smart enough not to stay in the cabbage patch.
2) Be smart enough not to fall asleep in the cabbage patch.
When I waited after those two, one child said plaintively, 'We can't think of any more.'

I said all right, we could go on to the third and last story. Then I mentioned that I thought the story said the opposite of what some of them had been trying to say. I thought it said something like this, 'It's OK to do what your mother tells you not to, as long as you don't get caught'.

They looked at me in mystification.

Story III

Same beginning as above. Once in the cabbage patch Peter Rabbit finds a treasure chest full of gold, takes it home, tells his mother where he found it, buys her a new dress, the family goes to Coney Island and has a ball. As they are going to bed, Mother Rabbit says, 'Isn't our Peter a wonderful little rabbit!'

This time there were a lot of hands with messages.
1) Finders keepers.
2) What people lose, you find, you keep.
3) Sometimes doing what your mother tells you not to do is OK because you might find money.

I asked the class if they would like to try to write stories with a message. They were eager. I asked if many wanted to use a fairy tale person and they did. So we listed some familiar names on the board — Cinderella, Three Bears, Rapunzel, etc. I said they could make their character do anything at all. If it helped them they could pick a message from those we had listed on the board. A couple of boys did not want to use fairy tale characters and I urged them to make up any character at all that they wanted to.

Some of the papers follow.

RAPUNZEL

Once a upon a time thar was a girl named Rapunzel. One day she went out for a walk. And a watch cate har. And she brat her to her Palars. And capet har for lase of yards. The wheal did not have stares. And Rapunzel had lage hegr, and she thake a big lather and put Rapunzel up it. And she said Rapunzel Rapunzel, lay down your hair so I will climb the golding stare. And that want on for yars and yars. One day the shch wate to the palles and said Rapunzel Rapunzel late down your hear. And a Prince saw the shee and saw Rapunzel and wan the sche want the prince said the sametime and wan he saw Rapunzel he kiss her.

When you see a wtch let her capture you.

[It would be interesting to compare the spelling in this fable with that of Chaucerian English.]

JUAN BOBO

One day Juan Bobo was looking for a job and he found one. His job was peeling banabas and they told him to peel the bananas and throw the cakasas away but then the bananas away the took the cacaras and them they hit him and threw him out of his job then he went to his mother house his mother was going out to the movies and he said o.k. and his mother told him when the baby cries give him milk and when the pig cry took put him out side and she went, he sat down and them the baby cry he forgot when they told him and put a injection and keeled the baby he told the slep but the baby was dead. And them the pig cry he forgot that she siad that the said o.k. pig I'll send you to your mother and you know what, he put all the gold purse on the pig and sent him a lone to the movie and he fell in a mud and was very dirty and when she came threw him out of the window and he went to a orange tree and ate 30 oranges and them ate 25 nuts, 30 pears and bananas and he drink whiskey and got to drunk, he climb a tree. The tree was about 300 feet and he touch he was in a swing and jumped and killed himself. So remember, don't eat to much.

* * *

Of course there are other ways to introduce the fable form. One can talk about the common character types in fables: the trickster, the hustler, the boaster. One can talk about the themes that occur in fables: the fate of the boaster, the consequences of lying, the triumph of the clever (or the stupid). One can read Thurber or Aesop, take stories out of magazines or newspapers and match them with morals, pick one fable and try to put a dozen inconsistent morals at the end and try to reinterpret the tale to make the moral fit.

There are any number of ways to begin, and when students do not feel judged or tested and when the theme or form they are shown can be related to some world or feeling that is familiar to them, getting young people to write or talk is not a problem. What is more difficult is to follow up on initial writing; to develop a sustained and comprehensive programme that enables the students to function without continual prodding or performing by the teacher.

One must find ways to make different kinds of imaginative writing lead to one another. In some classes, for example, fable writing led the students to write stories about origins. They wrote of the origins of different features and characteristics of man and the animals. These tales were close to fables and in two cases, on the origin of the camel, they were fables about origins.

THE CAMEL

Once upon a time there was a king who thought about a problem he faced. He wanted to go with the princess where ever she went, and she was going horse riding. He thought that if anything happened to her he would be there to protect her. He wanted to travel with the princess both on one horse without making her uncomfortable. He gathered together the wisest men in the country. They thought of making a saddle over a saddle but it didn't work. So the king said, 'What's wrong with a saddle behind a saddle?'' They tried it but it didn't fit the horse. The thought about having a special horse. The the king said, ''You can't make a horse like you make a house. They have to be bred.'' So they tried breeding the horses to come out with a horse for two. They came out with zebras, unicorns, antelopes, cows, deers, steers, and all sorts of animals. This is how these animals originated. The king was pleased with the other animals which were made by mistake, but he still wanted a horse for two. The king was bored and disgusted with this problem. The next day he had a wonderful idea. Once again the wise men were summoned to try out his idea. He wanted to know if two persons fit on one horse without the back rider falling out. He devised a method by changing the bone structure of the horse. This was done with a great mallet to bash in two spaces for the two persons after killing over 150 horses he was successful. He named his new animal the camel. As generations passed by the camel became more and more adopted to his new shape until he was finally developed. (The End)

THE WAY THE CAMEL BECAME

There once was a horse. He didn't want to be a horse, so he went to a witchdoctor. The witchdoctor said, "What do you want to be?" The horse said, "I want to be something with two humps on my back and a long neck." So the witch-doctor said O.K. but he was a fake. So when the horse was sleeping the witchdoctor took a hammer and hit him on the back and then humps came out. Then he pulled his neck and it was long. So the next morning the horse woke up and saw this and he said, "I don't want to be a camel. I want to be beautiful me." But he couldn't be helped.

Tales of the origins of things are indirect ways of talking about how the world got to the state it is in. One can imagine tales of the origins of war, love, fire, jealousy, greed, etc. Younger people seem to like using this form when writing for adults. They feel more at ease with indirection than with reasoned statements, and reveal more of themselves in tales of origins than they would dare to do in 'compositions'.

There are a number of different origin tales that can be explored following upon the writing of fables. For example, there are:

1. 'Why are ...?' tales, i.e. why are owls creatures of the night, why are men cruel to one another, why are there stars in the sky, why are there so many animals ...
2. 'Why is there ...?' tales, i.e. why is there hair, why two arms, five fingers, a heart and a mind, love, hate, confusion ...
3. 'Why do ...?' tales, i.e. why do men live in the air and fish in the water, why do animals let themselves be tamed by man, why do people stop growing ...
4. 'Why don't ...?' stories, i.e. why don't animals talk, why don't people fly ...

A good source of origin myths for the teacher is the *Larousse Encyclopedia of Mythology*.

Ballads and blues lyrics are fables in song, and one can move from writing fables to listening to them set to music. There are so many young people these days who write songs (not to mention those who listen to them eight hours a day) that to bring song lyrics into the classroom seems an important way to bring the life of the young into classrooms in a natural way. In this respect the teacher would do well to listen to and know the work of Bob Dylan, Joan Baez, Paul Simon, Carole King, Richie Havens, Al Green, Stevie Wonder, Aretha Franklin, Curtis Mayfield, Santana, Malo, Tower of Power, etc. Writing fables can lead to writing songs, to collecting lyrics, to listening to records and having the students transcribe the lyrics as they hear them or make up new verses to old songs.

In a sixth grade class I visited in New York the teacher made up a large number of cardboard discs that were the exact size and shape of 45 rpm records. The labels were blank except for the words:

There were two hooks on the wall. On one were the blank discs. On the other were discs where the students had filled in the name oftheir Number One song of the week. There was also a tally sheet on the wall

NAME OF SONG	NUMBER OF SELECTIONS	RANK

used to determine the order of the class's Top Ten for the week plus a Top Ten list,

NUMBER	SONG	GROUP	NUMBER LAST WEEK

as well as lyric sheets the students made up of all of the songs of the week as well as new verses written by the students. The teacher had created a mini-reading, writing and maths curriculum out of the music the students listened to outside school.

7

Some students will be less interested in fables, origin tales or song lyrics than in direct journalistic or technical writing. It is important to develop as diverse a writing programme as possible, one that encompasses group writing, exploring different forms such as the fable or parable or sonnet, etc., individual writing projects; novels, stories, poems, reports. There is no need to limit writing to the traditional English or 'language arts' curriculum.

Young people love to pass notes to each other, to communicate behind the teacher's back or despite the teacher. Note-writing is an obvious way to encourage students to express themselves on paper so long as the teacher can live with the idea that he or she does not have to read everything the students write, and that some writing is none of a teacher's business. Dan Peletz, a third grade teacher at Hillside Primary School in Berkeley, not only encouraged his students to write notes to each other, but gave each student a mimeograph master and suggested that they draw their own letterheads and make their own personalized stationery, which they did. Here is a sample:

Lew Hanson, an elementary school shop teacher in Berkeley, is one of the most effective reading and writing teachers I know. Everything in his shop, which can accommodate a dozen primary school children at a time, is labelled. Students have to write up order forms for tools and equipment they use, read plans and simplified how-to manuals. They are encouraged to come up with their own designs and write down instructions so that other children can make similar toys or games.

Sharon Nitta, a nutritionist and teacher at Hillside primary school in Berkeley, integrates writing into a cooking and diet planning curriculum she is developing in a kindergarten class. The children plan meals, write out and use recipes, and print them for the other classes. They are becoming involved in planning lunches for the whole school, producing menus and helping to buy, cook and serve the food. Writing is a small but essential part of the plan to turn as much responsibility as possible for maintaining and governing the school over to the students.

I have worked with a number of high school and junior high school students who were not interested in most of the forms of writing I presented, but who plunged into writing, printing, selling a series of how-to manuals they prepared on such diverse topics as:

How to fix your own bicycle
How to make dresses
How to fix a car
How to take care of a mini-bike
How to survive in the wilderness
How to cook like your grandmother
  Creole Style
  Yiddish Style
  Soul Style
  Mexican Style etc.

Selling these booklets was extremely important for the students who needed the money. Writing, which they had rejected along with everything else connected with school (even the open 'hip' school they attended), only came to make sense to them when they could turn it into cash.

8

I have discovered many teachers who listen to their pupils and make it possible for them to write honestly in the classroom. Spenser Jameson provides youngsters at Youth House in New York City, a prison school for youths awaiting court referral, with several hours a day of calm and serious work. I have observed the painful nervousness and disorientation of the pupils at Youth House, yet despite this Spenser Jameson's classroom transforms them. They sit and talk with their teacher about their problems and anxieties, and they are subtly led beyond the spoken word into art, music, and literature. M.J.B. discovered his poetic voice in his classroom in Youth House.

WHO

If there is someone
  who;
Cares for me and
watches over me with
an eye of protection
Please! ...
tell me
  who (?)

If it is you;
who has this
care — affection
or even love (!)
for me
Just
tell me it is you (?)

I ask for someone to
  come;
But all I do is
wait — but if I
ask you —
Please (!)
Will you
  come (?)

When I meet me
  death;
And there is no one
to weep for me,
And you think of me,
My plea;
  think of just the *good* me (?)

DREAM

In my sanity (when I possess it)
  no dreams are permitted
I can coagulate my thoughts with
  the utmost precision

Coordination is perfect and my reflexes
  stream with a new found adrenalin

I despise dreams (fantisy that is.)
  For children with their
Imense maturity dream.
  (People in society don't dream.)
I want to be important someday
  (similar to those in high society)

I wish I had the ability to
  Dream though —
(But people would say: there
  Goes that notorious man of society —
But he *dreams*)

THINGS

Tears —
  what are they?
  who needs them?
Emotions —
  I despise them!
Happiness —
  I'll never have it!
  Why should I see it?
These things —
  Please; I don't want
  any of them — so somebody
  take them away
  (I don't deserve them.)
Help —
  please somebody!
  (but I suppose I don't
  deserve that either.)

Others youngsters are not fortunate enough to have a teacher who takes them seriously. Charles Franklin, who attended the Baltimore public schools, writes poetry outside class, there being no place in the curriculum for what he has to express. Through a fortunate coincidence he showed a copy of his work to a teacher who seemed to be the sort of person who cared about poetry. This teacher passed Charles's poems to another teacher who showed them to me. This is the first time they have been published.

WAR BABIES

I saw them again last night,
Down by the railroad tracks.
Playing in the twisted rubble.
Sleeping under a bent ironing-board.
Bleeding in an abandoned truck.
Crying in a fresh-dugged grave.

I walked down by the sea,
And saw the little bodies floating
out with the tide.
Watched them stumbling down
the sand on broken stubs.
Looked in a cloud and saw them cough.
Screamed when they held their
arms to me.
Sobbed as they crawled away.
Cryed as I watched them leave.

They have no faces,
They have no feet.
They have no minds,
They have no hearts ...
They have no world.

ANOTHER MAN HAS DIED

Another man has died.
Give away his clothing to relatives,
Lying in wait beneath the oil furnace.
Under the cellar steps
Cowering from sight behind the
shower curtain.
Give his money away,
To the casketmaker;
To the priest;

To the garbageman;
To the slut in his bedroom.
Sell his house and his land
To the church ... looks good on
the record book.
Sell his children into slavery,
Claim his wife;
Fire up the ovens for his body,
And bury his soul in the ashes.

Another empire gone,
Buried in the dust of ancient bones.
Growing thin to the broken wings
of unfed doves.
Gleeming in the sweat of five
kings playing in a tomb.

Walk on the million universes that
the winds have torn from the desert ...
And urinate on the temple.
Another man burned on a cross,
One more soul floating in the
rising smoke of the ovens.
Five boys bleeding in an alley.
Ten boys running with red hands.

Watch the victims leave the world.
Follow them down to the graveyear,
And throw rocks at their mourners.
Spit on the caskets.

I guess I'll die again.
I do it so much now ... so much.
What a shame.
What a damn shame.

EXCERPT FROM: THE WORLD FROM A 3RD FLOOR APARTMENT

IX — "The circle"
Scream Gladys
alley beat

bare flat
bleedin' fat
Scream Gladys
my sometime Keeper
My
Church
Saved
Swinger
My hip-swingin'
mamma
Scream ...

At times things can happen outside school that compensate for closed, alienating classrooms. A summer programme in New York City brought together a teacher, Elaine Avidon, and a group of youngsters hungry to express themselves and weary of the emptiness of school. Together they created *What's Happening*, a magazine for 'the teenagers today to express what they feel'. The magazine was distributed free to youngsters, though adults had to pay.*

During its first year many youngsters found themselves writing instead of whispering and joking angrily in isolation.

THE BLACK MAN

I sleep at night
When the stars are bright,
And dream of the black man's
Flight through life.

I wake in the morn
Refreshed and fine,
But still the black man
Is on my mind.

Soon at noon
I come to know,
That the black man has
No place to go

*Unfortunately the magazine is no longer being published, and back numbers are hard to find.

By dinner time
My mind is clear,
To the fact that the black man
Was never here.
Dorothy Moody, aged 16

HEY NIGGER

Negro man,
Dark as can be
Negro man,
Father to me,
Nickname nigger,
No right whatsoever;
Color different
From any other;
White man must be much better!
He bombs my home,
Whatever I own;
Negro man,
What you go to do?
Sit still,
And let my son grow up like me
And you.
William Barbour, aged 16

IT'S A MIRACLE

Why was I created in this
World of frustration and damnation?

It's a miracle —

The addict on Lennox Ave, that thrives
On the $5 fix,
Who steals and kills
Just for kicks.

It's a miracle —

How man can construct and then destroy.

It's a miracle —

The White, Black, Yellow and Red
Inhabitants of this city and
This damned World.

The lush on Saturday, the Christian
On Sunday and back to the lush on
Monday.
How can these creatures exist;
The sacriligious, the religious
fanatic, the drug addict?
It's miraculous I tell you,
It's miraculous.

As I walk the streets of Manhattan
I sniff the contaminated air
Which was meant to smell like satin.

Where trees used to exist
Gigantic structures pop out.
Wherever you go.

Oh
I wish it wasn't so.
Heaven only knows
It's a miracle.

Why must I exist in this city
Or in this World of hate, sin
and destruction?
It's miraculous I tell you, It's down right miraculous.

Otto Grant, aged 15

There is no age at which a child starts to speak honestly or begins to conceal his thoughts. It depends upon whom he is speaking to and, too often, on what he feels they expect to hear. Deborah Meier, a pre-school teacher, spoke to her pupils and they drew for her, titled their pictures, told stories, and wrote poems. Some of their pictures and poems are on the following pages:

There was a lady
Who was walking along.
And a man dropped a hatchet
And it cut off her nose.
And her toes.

Jane Ellen, aged 4

A man hunt'n in a boat.
And he saw an egg fall out of the water.
The egg was cracked
And something was in the egg.
A baby bird!
And her mother came.
And saw the man holding the baby bird.
The mother bird land down on the man.
The mother kill the man.

Alaric, aged 4

There was a girl that
  jumped in the mud
Then she fell in the water
  and broke her leg.

She tried to get out
  and then the boy came.
And he fell in the water.

And the fish came and
  then the cat came.
And he jumped in the water.

And the lady came and
  that was their mother.
She jumped in too.

And the man came, and
  that was the stranger
And he and the dog came and
  jumped in the water.

The dragon ate the flag.

I am an Army man. I saw a rose.

And they all stayed there.
And slept in the water.

And that's all.

Joanie, aged 4

MY BROTHER

Paul did the wrong thing.
I did it with him too.
My mother was very mad.
Very mad.
We mixed up paints in a jar.
They were Paul's paints.
Red and yellow and blue and
white and pink and green.
My mother sent us into our
rooms.
"You can't come out til onne
hour passes."
We got some metals and melted it
into two swords.
You know what we did?
We scared my mother out of
the house with our swords.
... Then we made dinner.
... For our dad.
THE END

David W., aged 4

ABOUT ALARIC

A goat is much stronger than a giant.
He can knock down the house.
He can knock down a castle.
He can knock down a fire station.
He can knock down cars.
But not me.
I'm *much* stronger.
My daddy is a little stronger.
My baby brother is not stronger.
He can really talk like people now.

My brother Gregory is 3 years old.
I don't punch him.
He punches me.
I hug him.
He punches me because that makes
me cry.
It's not nice to hurt anyone.
I like to be nice.
But my baby brother doesn't.
My baby brother don't care.
I like Gregory, but I don't
know shy.
I like my friend Robbie cause he's nice.
He's nicer than grown-ups.
I want to be a cowboy.
Like a sheriff.
He puts bad people in jail.
A person can kill a bad person.
An Indian is a bad person.
A cat that chases mice is.
A dog that chases cats is.
A back robber is bad.
A jail makes him better.
Keep him in jail.

Alaric, aged 4

JOE'S STORY

My brother doesn't like me in
the morning.
But he likes me in the afternoon.
He likes me today.
After school I nap.
Then I fight with my little sister.
She's only 4. I'm only 5.
When she kicks me she runs away.
I catch her. I jump on her.
Then my mother comes. I go to Stephen's
house.
I'm kind of scared when fire engines come
at night.

I dream about them.
It's really scared.
The fire and the fire engine.
At monring I'm not scared ...
Joe, aged 4

MY FRIEND

I like to play with Alaric,
I like to play with Eldred.
He's a friend of mine who
lives in Madison Park and
goes to this school.
He's a Negro.
He's real big.
He's nice to play with.
We plays cowboys and Indians.
I get out my sisters and they
are the Indians and we are
the cowboys.
We think sometimes about how
we are going to play cowboys
and Indians.
Sometimes we play Pirates.
See, we make a treasure map
and we're pirates.
We go someplace and find
the treasure.
We dig with shovels.
We found bones of dinosaurs.
Way down deep.
Real way down deep.
I took them to the Museum.
That's how the dinosaur got
to the museum. It lived a
long, long time ago.
Robert, aged 4

9

Not everything that even the most committed and sensitive teachers try works with the students. For that reason it is important to try to understand what doesn't work as well as what does. However, it is more difficult to get at failures, and take criticisms as attack, rejection and accusations of incompetency.

When I first began teaching I was afraid to talk to other teachers and especially my supervisors about the things that were not working in my classroom. No one offered me any support and I felt that if my colleagues discovered that there was occasional chaos in my room, that the students were often bored, that I yelled and screamed and pleaded with them, then my brief teaching career would be over. No one ever told me that they had begun by teaching badly or that it was possible to teach oneself how to become effective. This is similar to the way that teachers fail to tell their pupils that it is natural to start out writing badly and possible to learn how to write well.

At the Teachers and Writers Collaborative our seminars were often concerned with analysing what didn't work. We all believed that it was possible for young people to be able to write with ease and skill and to look upon criticism not as rejection but as a way to avoid the same mistakes.

During our discussions of teaching writing fables, for example, we found that reading traditional fables as a way of getting young people to write fables of their own hardly ever worked. Surprisingly many teachers discovered that their students did not care to read fables and the best of the classical fables seemed irrelevant and academic to them. They were often put into the same category as any required reading material and were not read with any attention. The children were often deaf to the magic of the best fables until they tried to write their own. At that point they became interested in what other people did with the themes they addressed themselves to.

This same insight with respect to fable writing applies to other forms. Students who have written their own sestinas seem more likely to struggle with other writers' sestinas. The same is true for parables, haiku, limericks. I have found that to begin where the students are, introducing a new idea and letting them play with it, is a better strategy than to read some excellent examples of a form of writing and ask the students to try to write their own. Reading before writing seems to imply to the students that the examples are models to be imitated, not personal variations on a particular form. As soon as students worry over imitating great writers of literature they seem to lose their own voices and when they do write tend to come up with stale and laboured work.

There are other failures we all face as teachers. It is often difficult to interest students in material that one finds boring oneself. I don't care very much for some excellent contemporary verse. The language dazzles me in a way. I have heard poems which I do not myself like to read (though read by others they are clear and moving). I stay away from these works in the classroom.

Ideally there should be many different adults with a variety of literary sensibilities available to young people. Lacking that, a teacher might tell the students about the variety of writing that exists and about how sensibility develops out of the encounter of personal style, language and feeling with works created by others. Too many young people are left with the impression that there are some great works that must be loved by everyone and that if these works do not move them there is something deficient about their intelligence.

I respect the work of Joyce, Mann, Proust, Hesse and Virginia Woolf. At one time I was moved and excited by most of them. As I have reread them my views of them have become more complex and critical. I learned how to present works I didn't care for to my students over the last few years. Instead of dealing with the works directly I've always spent some time on the question of sensibility, preference and affinity and tried to bring my students to understand that there is nothing wrong with not being moved by what moves me or being moved by work that leaves me cold. We usually talk about musical preferences as a way of talking about style and sensibility. We listen to Mozart, Ornette Coleman, Verdi, to Schumann and Schubert songs and to Bob Dylan, Joan Baez, Aretha Franklin, the Impressions, Stevie Wonder. I try to overcome the moral battle over whose music is superior and replace it with considerations of who is most skilled within a given form (say lieder or folk ballad or rock) and how preferences develop.

There are other ways to get at the different affinities people develop toward particular forms of writing, singing, dancing, dressing and living.

To the ancients, all of nature was made up of four elements — earth, air, fire and water. These elements were the subject of scientific investigation; they were represented in paintings and described and invoked in poetry and music. Even today, all of us have associations with these elements and preference for one or other of them.

I begin by telling my students about the elements and the way in which some men felt they constituted the world. Then I ask the students to freely associate with each of the elements and put their associations on the board. Something like this usually results:

Earth: Warm, dirty, seed, brown, rich, green.
Air: Thin, pure, fly, high, colourless, open, head.
Fire: Heat, burn, hurt, passion, feel, love, die.
Water: Swim, cool, drown, float, clam, storm.

Then we talk about the associations, and about the differences that exist among the elements. After a while I ask the students to put down in order the elements they feel closest to. I do the same thing myself. Then we look at the class responses and see what our preferences are. The students look at each other's responses. There are no right and wrong answers — it is all a matter of personal preference. The elements simply tell the students things about themselves and each other.

Once a student suggested we make a chart of first preferences. It was interesting that most of the boys in the classroom preferred air and fire, and most of the girls preferred earth and water. For the Greeks air and fire were masculine elements, and earth and water feminine.

After we had examined everyone's preferences I asked the class to assume that they were each composed of some mixture of earth, air, fire and water and to make up charts of their own percentages. The students really began thinking about themselves and became curious about one another's choices. Later a student suggested that someone in the class leave the room and make up a chart about himself in the hall while everyone else made up their own charts on him. Then we could compare what the person thought of himself with how others perceived him.

I made no attempt to interpret what the students said about each other and themselves. There were no right or wrong responses, no marks involved. We moved from the elements into other areas where preferences are involved: colours, styles of dress, forms of writing, voice. After a while the students not only began to articulate their own literary preferences but began to be curious about each other's preferences and teach each other.

There is another failure that particularly distresses me. Some things that I find exciting bore my students. Despite my understanding of different taste I still want my students to be moved by poetry that moves me, to laugh at novels I find funny, to admire the authors I admire. It often doesn't happen and despite myself I hear a high moral tone slipping into my voice —

'You kids don't understand anything.'
'If only you'd pay attention ...'
etc.

But every time I have asked my students what they *did* like (if they couldn't like what I cared for), they have told me. Whenever I showed them respect and listened and learned about their worlds, the students returned the respect and listened to me.

There are other problems teachers run into. Some teachers insist that they have to read everything that is written in their classrooms and then find themselves limiting the writing done to fit the amount of time they have available for reading student papers. It is incredible how much written material can be turned out by twenty-five or thirty children in a week. I've often found myself overwhelmed by the amount of work to be read and wishing my students wouldn't be so prolific. Sometimes I've started slacking off, skimming stories or conveniently losing whole sets of writing exercises. Naturally the students get quite angry with me. The only way I have been able to deal with the problem is to explain to the class that I can only respond to one or two things per student per week and ask them to pick what they would like me to read and comment upon.

This solution provided the unexpected bonus of getting my students to judge their own work. Selecting papers for me to read and comment on became the occasion for self-criticism. Often the students picked their best work to show me. However, after a while some of them usually began to ask for comments on the works causing them the greatest problems — a story that ended badly, a poem that was flat, an essay that didn't express the writer's ideas.

I encourage students to bring out problems they are facing in writing and comment at length on their work without grading it. A grade seems too much like a final judgment and I want students to understand that they can repeatedly return to what they write, fixing and chopping and modifying until they come out with something that says what they intend in a voice they consider appropriate.

Commenting on and correcting student work is a sensitive matter since it is very easy in school to chase students away from writing. The best editors I have encountered in my career as a writer never tell me how to write and never take the licence to change my language. They ask pointed questions such as:

— do you mean this?

— the language is awkward — are you sure you want it to sound this way?

— doesn't the argument seem to get lost here?

— haven't you said this before on page 32?

— etc.

Editors understand that writers are usually paranoid about people tampering with their words and tread lightly though toughly. I have had articles returned to me with one or two questions about every paragraph. This usually means that I failed to write well and, though I feel put down at first, after a few days I come to deal with the questions.

The questioning form of responding can be combined with comments such as : this is a fine line, well put, etc. and other positive praise. However, not all comments need to be positive or merely questioning. When I first started teaching my tendency was not to give students negative criticism — never to say 'I don't like this story' or 'this is not one of your best' or 'you seem bored' — but I've learned that young people are quite tough about their own work and can take truthful comment if they trust you.

In a setting where a lot of writing is going on all the time each person is bound to produce good and bad work; to praise everything equally is a form of deceit. For some students who are just writing for the first time support has to take precedence over criticism. There is no formula for telling a teacher how to balance questions, praise, and direct criticism. However, it is best not to lie and say you like a piece of work that you don't. That might mean occasionally shifting your comments from the quality of the work to the nature of the author's effort. I tell students that they are beginning to get where they want to go, that they put a lot of work into something; and encourage them to keep on writing. Once fluency and ease (which form the daily confident habit of writing) are developed, it is easier to begin to respond to a student's work in depth and help him or her develop the coordinates of self-criticism.

People often wonder about how spelling and punctuation are dealt with in an open-ended writing programme where the teacher does not always correct papers. It seems to me that a teacher should make it clear that he is available to correct any errors and to help the student re-work his ideas or restructure his sentences, but he should not impose correction. This does not mean that student writing will remain as awkward and ungrammatical as it may start out being. On the contrary, *students want to get things right* and will ask for help if it is not forced upon them. I have found that students learn more about the need for uniform spelling and good grammar by reading each other's works and discovering those points at which grammar and spelling are necessary for communication, than by reading any grammar books or doing any number of exercises.

Writing ought to be done as often as possible without its becoming boring. In my classroom we wrote a collective theme once a day. I would set the theme or one of the students would set it. Often the students would take their themes from TV commercials, which they learned to parody. We wrote, for example, about Marlboro country, about getting relief quicker, about aches and pains. No-Doz and wake-up, Sun-Up and tone-down. American advertising presents a wealth of material to be used, varied, parodied and played with in the classroom.

As the year develops, students can take increasingly active roles in the development of writing themes. Often in my class all that the collective writing for the day consisted of was a joke on a common theme, a fable or a snatch of conversation. I wrote with my students and found that I faced the same anxiety and periodic blankness that my students faced. It made it easier for me to understand their writing and for them to try bold things, since they saw from my writing that I wasn't perfect and therefore couldn't reasonably expect perfection from them.

Many of the themes produced in my classroom were read out loud and then forgotten. No student did all of the themes. Some just couldn't write jokes, others couldn't write fables. That's natural since we do not all speak with the same voice. I commented on some but not all of the papers, and reproduced some. I found that it was not necessary to treat all assignments in the same way and with the same intensity, and this increased my freedom as a teacher.

Along with the collective writing exercises I tried to individualize writing. Whenever a student mentioned an interesting theme or developed a character or story or idea that had promise I suggested he treat it at length, and gave him a notebook for his work. Writing also became a part of the science and mathematics programme. Once one realizes that writing is not merely a skill to be acquired but an activity that can pervade all the other activities of school life, anything can develop.

10

Sometimes it is difficult to see the potential in a child's writing. I once had a young Puerto Rican boy in a sixth grade class. He was shy and, according to the record, had an IQ of 79 and was illiterate. He listened intensely in class when I taught reading, otherwise he seemed to be somewhere else. He never spoke in class, yet after the Christmas holiday he came to me and told me that I had taught him how to read. It seemed that the idea that words were divided into syllables excited him, and so over the Christmas vacation he divided all the names under 'A' in the phone book into syllables and learned how to read. I was astonished at his excitement over a fact of grammar that seemed dull and matter-of-fact to me. I encouraged Carlos to write, and for all his struggle with English language, a beautiful, sad world emerged:

> One cold rainy day I was going to school and I had to go 1,000 miles to get there and there wasn't no cars and no buses and train so I had to walk. I got soke a wet. I still had 500 more miles to go at last I almost got there and went I got there the school was close and I thought for a minute and then I remember it was a haliday and then I droped deid.

> It rains too much and my flowers vegetable and gardens they get too much water. I got to think of something fast because if it keeps on like this my plants can't grow. So one day I was walking in the street when I saw this store selling rain supplies so I went in and got some then I went back home and I had one that will just rite rain so I planted in the ground and the next day I couldn't believe my eyes all the plants were just growing up. So I live happily ever after.

> I just don't like to think because every time I think I get a headache because one time I was thinking about the world fair and I build a mental picture in my mind I was enjoying myself then I stop thinking. I was going home went suddenly I felt something in my mind and I got a headache and I was criing because my mind hurt. From that day on I can't think.

> It happens every time I go to bed I forget to brush your teeth. then the second time I forget to brush your teeth. Then the second time I forgot the third time I forgot too so

I had to do something, so one day I was very sleepy I was going to my bed then suddenly I open my eyes then I remember and I ran back to bath tob and I brush my teeth you didn't got me this time so I went back to bed and then every single day I brush my teeth live happly ever after.

Carlos, aged 12

I gave Carlos books to read and encouraged him to listen to the people around him and record their speech and manner, and try to understand and capture their style. Trying to help Carlos I found there were kinds of heightened perception that supposedly 'limited' children can develop as enthusiastically as they evolve their own styles of life. I feel that there are many unexplored possibilities of developing writing and perception, perhaps for the not-so-sensitive as well as for the hypersensitive.

Carlos still tries to write in a half-hearted manner, but he receives no encouragement. It seems that writing is not a subject given high priority at the junior high school he attends. It is more concerned to teach mathematics and science to those few children who will make it.

Up to this point I have been primarily concerned with the writing that can be taught all children. Any significant programme must also provide for those few children who not only write but write with a seriousness and intensity that is not usual. These children, even at the age of twelve, would like to become writers, and in a few cases there are children whose work shows such a stamp of individuality and such unmistakable perception and love of language that there is hope that if they are taken seriously they could become writers.

Such children pose special problems for the schools because they may not conform to the usual measure of academic success. They are not likely to be excellent scientists or mathematicians and are more than likely to be individualistic, somewhat withdrawn and self-sufficient. Most teachers do not know what to do with these children and are usually content to let them sit silently in their classes. If their talents are discovered they are rarely made much of since they are not likely to fit into the rigid categories that the teacher judges as excellent.

The story of the last child I would like to introduce here is one of a sad and continuing waste of human talent. Louis, a boy of eleven, passionately loved language, and whenever he was not overburdened with his personal problems he played with language with unusual mastery. I never had a pupil who absorbed words more quickly and intuitively than Louis. Nor have I had one who was so discontented with static, colourless style, or so articulate in criticizing the banal readers that are provided in the schools.

When I took Louis to Cacoyannis's *Elektra* he said, 'You mean that a myth is a way of saying something about yourself, something that other people could understand and would be better than us, more beautiful; it wouldn't hurt me so much to tell it that way.'

Foreword

This story called Elektra is of the deepest passion and the deepest hope of avengence of her father's death. Her father was called Aggememnon, Aggemomnom was the rightful ruler of Argos ... He had been crudely slaughtered by his wife Clytemnestra and her lover Aegisthus.

Louis never finished *Elektra*. Nor did he finish *The Boy in the Slums*, still one of his family's prized possessions.

THE BOY IN THE SLUMS

Foreword

This story is about a boy namely me, who live in an apartment in and around the slum area. I feel that other people should be interested in what I have to say and just like me, TRY to do something about it, either by literal or diatribe means. This book is only to be read by men and women boys and girls who feel deeply serious about segregation and feel that this is no joke. Especially when you are younger you have a better opportunity to speak about and be willing to work for these problems of the slums.

1 — Do you live in the slums?

2 — How do you think you would feel if you did?

3 — Would you rather be rich have maids and servants to take care of you while your mother is away to dinners,

nightclubs, and business trips? Or would you rather be poor and your mother'd be home to *Love* and take care of you?

Before I wrote the last question down I made sure that at least *I* knew the answer. I had a decision to make also because my mother asked me that same question just a few days ago and take it from me its not easy to answer a question like that. But if just by mere curiosity you would like to know my answer to this question just open the pages of this book and read to your hearts content and do me a favor (just as a friend) tell other people about this book and *maybe* they may be encouraged to read this Book. (Oh, by the way all through this book a word will be underlined and if by any chance you want to know what this word means just look it up in the back of this Book it is called "Louis slang Dictionary")

I — AN INTRODUCTION TO MY MOTHER

I am dreaming and crying in my sleep.

I am dreaming because I have nothing better to do and crying because I am dreaming about a problem I had in school, you see I promise myself I'll be good and try to learn more, but everytime I come into the classromm (in my dream) my teacher right then and there starts to pick on me Louis this or Louis that. So I say to myself "Enough is too much, everyday the same old problem" why that's enough even to make a laughing hyena cry (wouldn't you if you were in my situation?)

Just as I was about to cry in my sleep for the second time unexpectedly a hand hit me right on my rear end (I knew it was a hand because I had felt this more than once). Of course I woke up and immediately knew that it was time for my brothers and my sisters and me to get ready to go to school. My youngest sisters name is Rene she is 3 years next comes Pamela she is 8 years old, my next sister's name is Alice. She is 9 years old, then comes my Brother who's name is Robert he is 10, then come me Louis I am 11, then comes my next sister Diane she is 12 years going on 13, and last but not least My Oldest Sister who's name is Barbara she is 14. I know you're not interested in my private life but

I'll fill you in a little way just to have something interesting to say. The first thing I have to do is head straight for the Bathe-room (PS By the way the word Bathe is just a fancy word I picked up from my teacher "Mr. Herbert Kohl". You know I'll let you in on a little secret, Mr. Kohl is kind of fancy himself. The reason why I'm telling you this is because my teacher told me to express myself to the *fullest extent* (that's another fancy word I learned from my teacher)

And the first thing I do in the Bathe room is to wash my face and comb my hair while my mother is ironing my shirt and pants. Oh by the way my mother's name is Mrs. Helen Frost (you can call her *ma* or Mrs. Helen cause that what I always call her and she doesn't get mad either). The next thing I do is eat my breakfast which consists of two or three jelly sandwiches and a glass of water or if I'm lucky I'll have a bowl of cereal with *can milk*. At this time it should be 8:30 time to go to school. P.S. 79 here I come I say as I start out of the door to my building. As I walk to the school which is within walking distance from my house I begin to think of things that could but then again couldn't happen. For example: (Maybe someda, I'll be a scientist or a big business man or maybe even an engineer or then again the President of the United States or maybe even the mayor. As long as it is somebody important. You see! some people are lucky enough to be born important but not me I'll have to work my way up to what I want to be (my personal opinion of the situation!) if I'm even lucky enough to get that far up as a matter of fact I'll even be lucky if I get past the sixth grade the way things are going now. If you ever get into a situation similar to mine take my advise don't give up you have to work for your goal, don't worry you'll never be alone in your problems other people just like you are sharing your same problems.

I feel I have to close this chapter now for I'm digging into my long buried problems which you probably wouldn't be interested in anyhow. But do me a favor, read on to the next chapter.

At present Louis is nowhere. His interest in writing is useless in school and his sensitivity to words useless on the streets. He believes in himself and refuses to yield his pride, even if it drives him out of school. The guidance counsellor of his junior high school insisted he take a vocational course in high school, but Louis persisted and is now in an academic course in high school. His teachers feel it is a mistake. He laughs and has taken an Arabic name, an 'original' name, a name that is his strength. His Muslim commitment is not out of hatred — it is a sign of the pride and self-love he has been able to preserve in an unbelieving and hating world.

It is hard to believe that this is necessary or inevitable. Teachers must be taught to look for sensibility and feeling in their pupils, as well as the abilities to perform intellectual tasks. Children's literature involves experiment and play. Teachers usually try too hard to interpret their pupil's work. If a child writes about violence he is looked upon as expressing violent impulses that are 'really' within him. If he writes about loneliness his teacher tries to provide him with companionship. This usual view of writing condescendingly implies that the child is incapable of literary exploration. Worse, it implies he is as humourless as the adults who assume responsibility for his education. I have laughed, cried, been duped, outraged and sometimes bored by what my pupils have written — and I have told them this. Their effort to understand themselves and the things around them demands no less.

Recommended reading

Following is a list of books I have found to be valuable practical resources for planning writing projects, some of them available in the UK.*

The Whole World Catalogue is the best collection of ideas, themes, materials and strategies for developing writing I have seen. It can be obtained from the Teachers and Writers Collaborative, P.S. 3, 490 Hudson Street, New York, NY 10014, $1.50.

*given without prices if out of print.

The Teachers and Writers Collaborative also has available two other valuable curriculum units and a newsletter full of practical ideas and diary accounts of particular classroom experiences: *The Teachers and Writers Collaborative Newsletter* ($2.50, four issues for $4.00).

A day dream I had at night by Roger Landrum and Children from PS 1 and PS 42 (Teachers and Writers Collaborative, Summer, 1971, $1.00).

Imaginary Worlds by Richard Murphy (Teachers and Writers Collaborative, Summer, 1971, $1.00).

Some interesting essays on teaching writing, written primarily by writers who also happen to be teachers, and which emphasize the process of writing (as opposed to ideas or materials) are:

Somebody Turned on a Tap in These Kids, ed. Nancy Larrick (Delacorte Press, 1971, $5.95).

Writers as Teachers, Teachers as Writers, ed. Jonathan Baumabach (Holt, Reinhart and Winston, 1970, £1.55).

Some workbooks provide interesting exercises in metaphor and fantasy. I have used the two mentioned here quite selectively, pulling out a page here, an idea there. Other teachers I know use the whole books.

Making It Strange, Books 1, 2 and 3, prepared by Sunectics Inc (Harper and Row, 1968, £1.70).

Experiential English, Sandra Fluck (Glencoe Press, 1973, $5.95).

There are many collections of writing by young people available these days, such as:

The Voice of the Children, collected by June Jordan and Terri Bush (Holt, Reinhart and Winston, 1970, £1.25 in paperback).

The Best of 40 Acres, ed. Cyril James (40 Acres and a Mule, New York, 1972).

Miracles, ed. Richard Lewis (Simon and Schuster, 1966, $5.95).

Journeys, ed. Richard Lewis (Simon and Schuster, 1969, $4.95).

There are Two Lives, ed. Richard Lewis (Simon and Schuster, 1970, $4.95).

This New Season, ed. Chris Searle (Caldar and Boyars, 1973,

£1.25 paperback): a collection of his students' poetry, prose, plays and drawings.

David Holbrook has produced a number of interesting psycho-analytically-oriented books on young people's writing. The books are full of samples of writing produced in classrooms in England (Holbrook's own especially). In some cases the writing is analysed in some depth and with great sensitivity. I find Holbrook's psychological analyses quite interesting, especially since they deal with young people's writing in a different way from most American works on the subject. The most interesting of Holbrook's books are:

The Secret Places (Methuen, 1964).

English for the Rejected (Cambridge University Press, 1965, £2.30 in paperback).

Children's Writing: A Sampler for Student Teachers (Cambridge University Press, 1967, £4.60; paper £1.95).

Kenneth Koch's two books are very useful and practical. The first deals with getting young people to write poetry. The other, which is excellent, shows how classical poetry can be taught in a natural and interesting way and be combined with student writing.

Wishes, Lies and Dreams (Random House, 1970, $7.95; paper $1.95).

Rose, Where Did You Get that Red? (Random House, 1973, $7.95).

Ken MacRorie has developed interesting and funny ways to get college students to write naturally despite the damage they might have suffered in high school and the early grades. Many of his ideas can be reinterpreted for use with younger people.

Uptaught (Haydon Book Company, 1970, $2.50).

Telling Writing (Haydon Book Company, 1971, $4.45).

Two useful general books are:

Understanding Children Writing by Carol Burgess *et al* (Penguin, 1973, 50p).

Children and Stories, a study of children's story-making from playground gossip to literature, by Anthony Jones and June Buttrey (Blackwell, 1971, £2.50).

Finally, for those people interested in integrating drama, improvisation and writing:

Improvisation for the Theater by Viola Spolin (Northwestern

University Press, 1963, $10.00; $7.50, text ed.).
Creative Playmaking in the Primary School by Paul Cornwell (Granada, 1970, £1.10).
Drama Casebook, John Challen (Methuen, 1973, £2.90, paper £1.50).

2 Games and Maths

I go to toy stores the way other people go to bars, bookstores, garage sales or thrift shops. Several times a week I wander through toy stores in Berkeley looking at games and models and dolls, reading instructions and experimenting with the toys whenever the salespeople allow me to. I love observing children deciding upon birthday presents or cajoling adults into buying things for them. I also watch parents trying to convince children to buy what they as adults feel they missed as children. The children's resistance to their advice is usually serious and 'grown-up' — they insist that toys are their business.

From toy stores one can get many teaching ideas, for they make clear what interests young people outside school. This doesn't mean that commercial toys are fine and sensibly designed learning materials. On the contrary they often embody the worst in our culture — violence, competitiveness, sexism, racism and deceit. Most toys are packaged to seem grander than they are. Everything is displayed in boxes that are twice as large as necessary on the assumption that the bigger the box the more people can be tricked into paying for it. Yet there still remains so much of experimentation, fantasy, love and the sheer pleasure of learning in games and toys that they draw children to them.

Certainly toys have a powerful effect in training children to take their place in the adult world. GI Joe is designed to have children practise violence, or at least to admire uniforms and get accustomed to technological warfare. It is no accident that laser guns, space modules, electronic bugs and missile systems are sold as birthday presents for young children. However, toys lend themselves to many uses and often the intent of the adults making the toys is thwarted by the ingenuity of children and by their repugnance toward some of the more dehumanizing aspects of our culture. Kids have to be taught to observe the habits of their elders and unless they admire what the adults do they often have to be forced to do so. Imitation of adult culture is not to be taken for granted, especially in cultures like ours where so many adults seem so often discontented.

I do not like Barbie dolls or what they represent. The dolls themselves are not so expensive. But they are surrounded by possessions that are meant to mobilize the greediest sentiments of young children. In the Barbie section of a toy store one can find Barbie clothes, Barbie wigs, Barbie cosmetics, a Barbie house and even a Barbie camper complete with sleeping bag and camp stools. Still, my daughters love Barbie dolls. I broke down last Christmas and got Erica a Barbie camper. The camper is central to many of my daughters' most elaborate fantasies and games. It has been everything from an actual camper to an ice cream truck to a mystery car carrying superheroes and heroines to a boat, a plane, a raft. Barbie is Barbie one day, Wonder Woman the next. My wife and I don't buy Barbie's elaborate costumes so the children create their own. The other night the Barbies in our house even had a political meeting on the top of the camper to decide how to create kid power.

It is probably true that there are children in this country who believe there is only one way to play with Barbie dolls. However, most children know how to discard the rules and create their own.

I have seen dozens of versions of chess, checkers, Monopoly created by children; seen musical instruments played in the most unexpected ways; seen children merge dinosaurs, stuffed animals, Barbies and knights and soldiers as they experiment with all the combinations and relationships they can imagine.

Watching children play, I decided that if one presents young children with the components of games, they will generate games themselves. Children experiment with different ways of doing things whereas adults get accustomed to believing there is one right and one wrong way to do things. Creating a game is much like discovering how to write in one's own voice. Making games, writing, building are all ways young people can discover that they can put things into the world, that they can have some control over life.

The other day I ran into some children who were playing their own version of chess. The knights jumped two squares at a time, since they galloped like horses. The queen, rooks, pawns and bishops moved in their regular ways, though the king was given the mobility of a queen. When I came upon the game a young student teacher was telling the kids how wrong they were in daring to change the rules of the game. He pointed out that there was only one way to play the game, that they would never be 'real' chess players if they didn't play by the rules. One of the kids said that she didn't want to be a real chess player but was curious about what happened when you changed the rules. The new game was interesting but the student teacher insisted it wasn't a game and forced the kids to play by the standard rules. When I mentioned to him that there were dozens of variations of chess played throughout the world, he claimed that the only justification for letting children play games in school was to accustom them to learning to play by the rules. I disagree. It is important to learn how to play by other people's rules but it is also important for young people to understand that all rules are not sacred or universal, that they can often be modified or discarded or replaced. Creating new games or modifying old ones is one way young people can explore things for themselves. It is an indirect yet powerful way of accustoming people to build for themselves and not to accept traditional forms as inevitable and timeless.

For the past few years I've been experimenting with ways of teaching games, creating new games and modifying old ones. During this time I have developed a scheme which breaks most games down into different components which can be modified and examined in some depth. Rather than beginning with specific games and showing how they have been dealt with in particular learning situations, I will start with some general aspects of games and then show how individual teachers and students have dealt with them.

From my perspective games can be examined according to:

1. their *themes*, the ideas they embody or the images that give rise to them;

2. the *playing board*, the space used to play the game, the environment specifically created for the game;

3. *the pieces*, the tokens or elements out of which the game is built, as well as the *moves* assigned to these pieces and the *captures* or *changes in status* that occur during the game.

Related to the moves are

4. the *decision devices* (such as dice and spinners) that are sometimes used to determine specific moves.

Overriding all of these considerations are

5. the *game's goals*. Is there to be a winner or loser, or are collective goals set? Does one need points or have to trap

an opponent's pieces or clear the board? How is the game scored?

Finally,

6. *teaching the game.* How do people learn to play games, or teach others to play? Which teaching strategies work and which fail to get students involved in games? How would a games centre in a classroom look? What kinds of skills and learning are involved in playing games and how can these be evaluated and documented? How could games be introduced into all areas of the curriculum? What are the resources a teacher can draw on and the materials he or she ought to have available to begin a games programme?

I Game themes

Themes suggest games. It is easy to imagine games centred about any of the following subjects:

— founding a city
— borders and boundaries
— bees
— pollution of a stream, a city, a park
— confrontation
— collective action
— dinosaurs
— rebuilding a devastated world
— the solar system
— discovering a new planet
— the meeting of two cultures that previously had no contact
— pigeons and people
— advertising a product in order to seduce people into buying what they don't need
— making it in a hierarchical system
— life in a mental institution
— life in prison
— war
— falling in love

— having children.

The themes that can be worked into games are inexhaustible. However, once a theme is chosen there are a number of different ways to develop a game, depending upon whether the game is the main reason for considering the theme or just a part of a larger programme. For example, many four-, five- and six-year-olds are obsessed with prehistoric animals. They love to play with plastic dinosaurs and in fantasy assume the size and power the creatures are supposed to have had. I think the fact that dinosaurs are extinct helps also, since real creatures could easily be lurking in a closet or in the woods or around the corner. Powerful extinct animals provide an ideal world for young children to experiment with power and violence and conflict.

In the Early Learning Center, a public alternative primary school in Berkeley, dinosaurs were the rage for several months. A number of the children memorized the names of all the dinosaurs pictured in their books. They made dinosaur costumes, papier-mâché dinosaurs, clay dinosaurs — they drew pictures of dinosaurs, told stories about them, put together plastic models, put on plays. They also made games. These games were just a part of their temporary obsession with prehistoric animals yet provided a way for the children to analyse and sort out a lot of ideas that developed in their fantasies as well as acquire and use a lot of factual information about the animals and imagine the world inhabited by them.

Several of these children came to my house one day to play with plastic dinosaurs. On our dining room wall is a large poster which lists twenty prehistoric animals. Under each picture is a brief description of the animal. Under the picture of the giant brontosaurus it says:

> The 'Thunder Lizard' was one of the larger dinosaurs, measuring about 70 feet long. Peaceful and slow moving, he spent most of his time in marshes and streams. His long neck enabled him to feed on underwater plants.

The kids asked me to read all of the descriptions over and over, and then picked up a checker board and began to work out a dinosaur game. They had plastic sabre-toothed tigers, tyrannosauri, brontosauri and stegosauri. They talked about the different qualities and powers of each of the dinosaurs, created moves for them and played some game in which the animals moved about the board and captured each other occasionally. I was curious about what the children thought they were doing and tried to get them to talk about the game they made. Each time I asked questions the kids withdrew. My analysis was destroying the fun of the game.

The children invented other dinosaur games. They made a simple board:

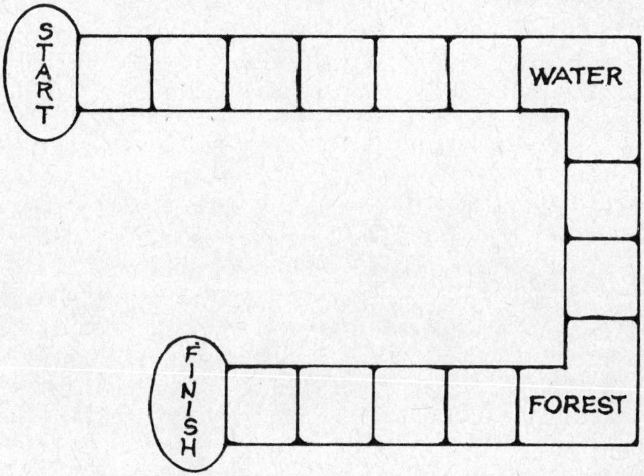

Each player chose one of four dinosaurs and a single die was used to determine the number of squares to move per turn. If a brontosaurus or plesiosaurus landed in the water the player got an extra turn since they were water creatures. If the tyrannosaurus or stegosaurus, land animals, ended up there, they lost a turn. The opposite happened when the players landed in a forest. The first person to get to the finish won. The game was extremely simple. The children who made it up taught it to others and for a few weeks it was played obsessively. After a while it became boring and was put aside. However, the children who played the game learned to read the words 'tyrannosaurus', 'stegosaurus', 'brontosaurus' and 'plesiosaurus', as well as 'start', 'water', 'forest' and 'finish'. They discovered that there was a 'saurus' in all the dinosaur names and asked what a 'saurus' was. I looked the word up in

Eric Partridge's *Origins*, an excellent etymological dictionary, and explained that it meant lizard-like. I also explained that 'dino-' meant terrible and therefore that dinosaur referred to lizard-like creatures that people imagined were terrible and terrifying.

The dinosaur words were more complex than the words the children were exposed to in school. They learned them with ease, however, since they weren't drilled or tested on the words but were free to play with them.

Games can often assume the same functions that drill does in more traditional learning situations. They provide occasions for repetition and therefore enable children to master certain academic skills in a setting where learning is combined with play.

Recently I proposed to a group of high school students that we create a series of games that focused on bees. During our first talk a lot of questions arose: What kinds of bees exist? Who are their enemies? How is honey produced? Do bees compete with each other? Do people get stung when they try to get honey? None of us knew all the answers to these questions so instead of making games we began to study bees. A few visited a friend who raises bees, I read about bees' communication through dance, and we broke down the structure of the bee community and began to sketch possible game boards. It was immediately clear that the cells on the game board should be the same shape as the cells in a hive, so we drew a series of hexagons:

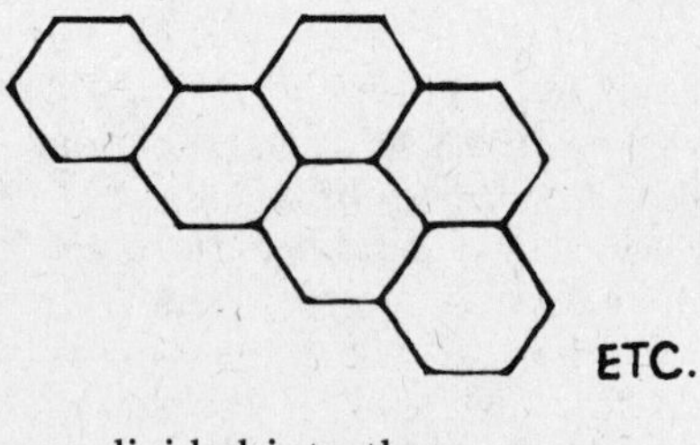

Then the board was divided into three areas:

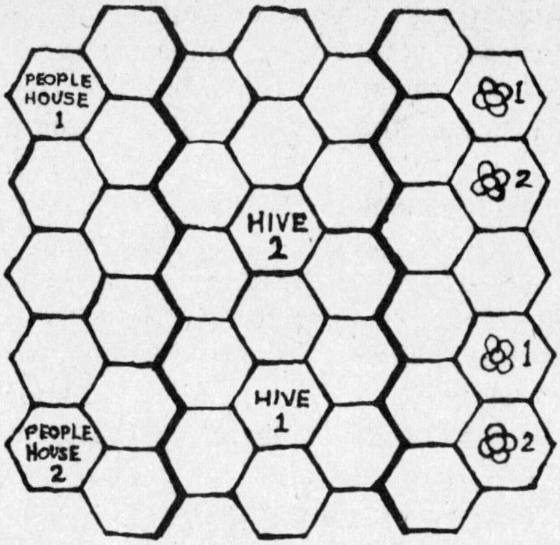

Then a first, simplified game developed for two players. Player 1 had hive 1 and house 1; player 2 hive 2 and house 2. In each hive was a single bee, B1 and B2. In each house was a person, P1, P2. The object of the game was to have the bees visit the flowers and return to the hive with pollen, and then for the person to visit the hive and bring the honey home. Bee 1 could only visit flowers marked 1 and Bee 2 could only visit flowers marked 2. Each player took turns moving across the board divided into hexagons and the pieces could move two hexagons at a turn. There was no simple direct route to the honey or the house because hive 1 was opposite house 2 and vice versa, and since the flowers were arranged 1,2,1,2, the pieces could also block each other. Some interesting strategies therefore developed.

We played this game until simple ways of winning or tying became apparent. More complex bee games involving workers, drones and the queen bee are being worked out. Some people are more interested in bees than in bee games, while some of the students are interested in neither. However, everyone seems to feel that the exploration opened up some techniques that could be used in the analysis of almost any subject.

For example, setting up a game and then playing it involve continual experimentation with the rules. Often problem situations arise which were not originally expected and therefore the rules have to be modified. Some games turn out to be boring. Others involve a simple strategy which once known guarantees one of the players a win, making the game uninteresting once the strategy has been uncovered.

Exploring a game and then modifying its rules has many similarities with the way theories are constructed in the sciences and social sciences. A theory is generally created to explain or provide a model of some phenomenon. Then the theory is explored. Experiments are performed to see if it is adequate; modifications are made; the consistency of the theory is examined. Scientific work often depends on playing with rules, and play with games can be good practice in scientific thinking for young people. Creating and playing a game is one way to become familiar with an unfamiliar idea and to understand some of the nuances of life systems.

Some of the other subjects that have been explored through game making combined with research are predator-prey relationships and life cycles in controlled environments such as aquariums and terrariums.

Recently I have become interested in the theme of boundaries. I've studied maps and globes and watched land surveyors work, looking at the way they mark out subdivisions and identify plot lines. So far I've just begun to turn these observations into games. Starting with a piece of graph paper it is possible to identify different boundaries:

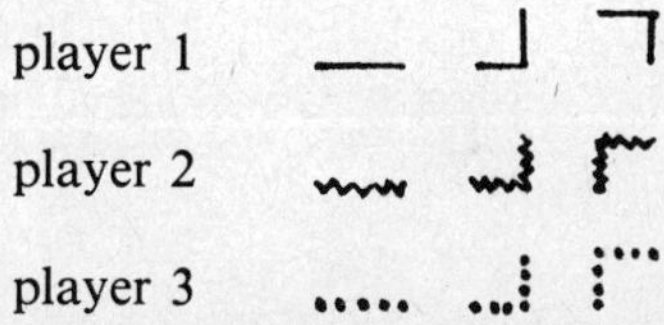

In a simple game, players would try to carve out spaces of their own on a defined grid, given the constraint that they can only move two lengths of a square (i.e. —— or ┘ or ┐ etc.) per turn. A simple version might look like this:

first set
of moves:

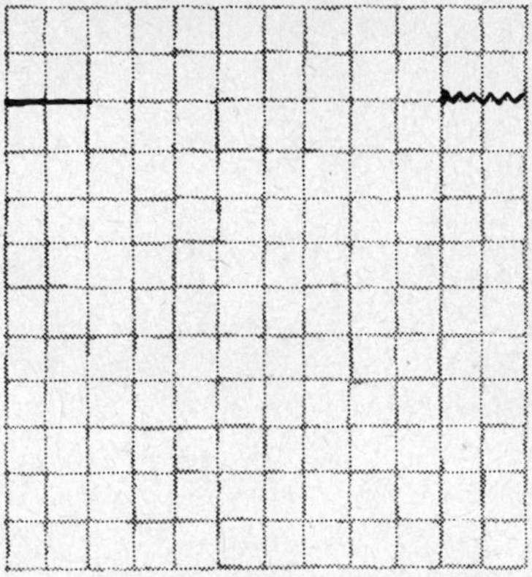

second set
of moves:

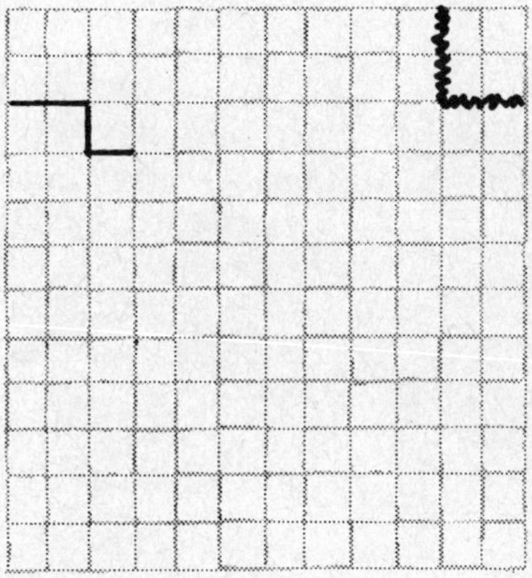

the end of the game might look like this:

last move:

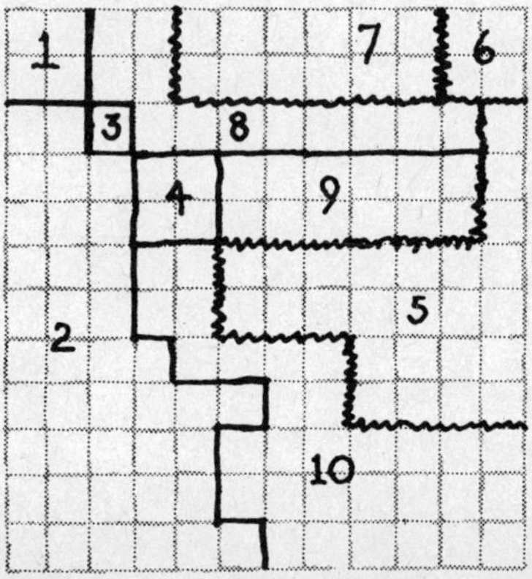

with areas 1, 2, 3, 4 being won by player 1;

areas 5, 6, 7 being won by player 2;

and areas 8, 9, 10 being won by no one since they are bounded by both players.

The game can be scored by the number of squares included in each player's boundary so that:

player 1	area 1	4 squares
	area 2	40 squares
	area 3	1 square
	area 4	4 squares
total		49 squares
player 2	area 5	25 squares
	area 6	4 squares
	area 7	12 squares
total		41 squares

so that player 1 would win by a score of 49 to 41.*

There are other ways to score the game however. For example, the winner could be the player who surrounded the largest number of areas rather than the most squares. In this case player 1 would still be the winner since he or she has surrounded four areas (1, 2, 3, 4) to player 2's three areas (5, 6, 7).

Boundary games tend to be more complex. For example, keeping the same rules for moving, the board can be changed to look like this:

* This game is similar to the classical Japanese game of Go.

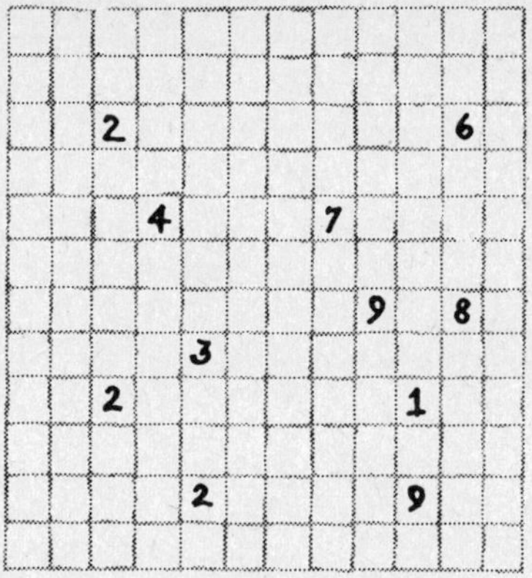

with the numbers representing bonus points to be gotten for surrounding the square containing the number. This game probably could substitute for a great deal of drill with basic addition facts in the early grades.

Another modification of the board could be:

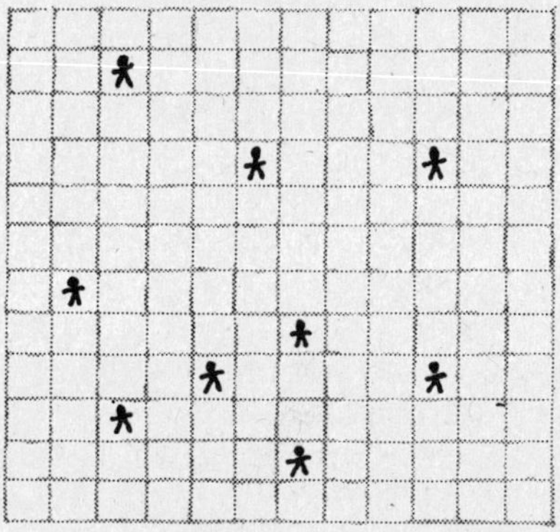

with the object of the game being to surround as many people as possible.

A few weeks ago I was trying to explain to some friends how different themes can be worked into games. They challenged me to turn the theme of having children into a game. Here is the result, which hasn't been played yet but might turn out to be an interesting game.

Obviously there have to be two different kinds of pieces — women ♀ and men ♂. I imagined that there would be two people playing, each having four men and four women on their side so that the game would have sixteen pieces:

player 1

player 2

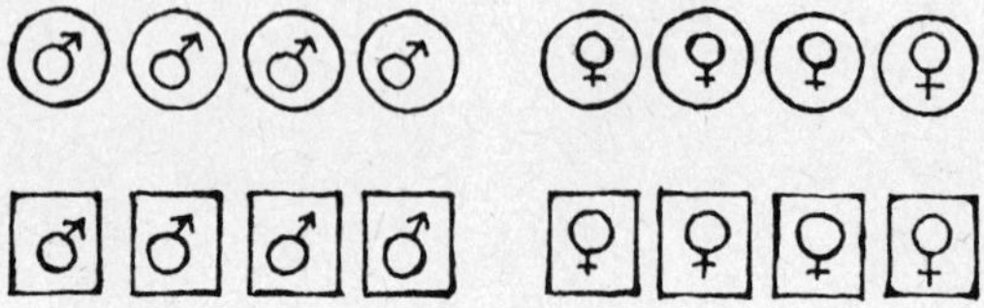

If a man of one side landed on a woman of the same side or a woman landed on a man there was a chance they would have a child. I made up a spinner to be used to determine whether a child would be conceived and what sex it would be. The spinner looked like this:

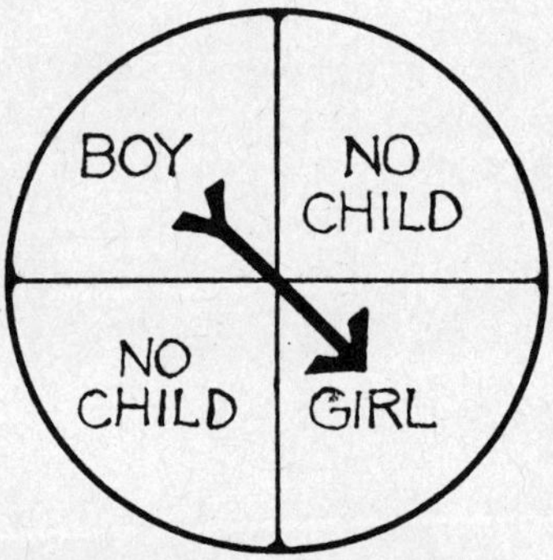

Then I made a board and set up a few more rules. Men can move one or three spaces at a time in any direction. Woman can move two or four squares at a time in any direction. If you land on an opponent's piece you capture it and take it off the board. If you land on your own piece of the other sex you spin the spinner to discover the issue of the union. If a child is born you add a piece of that sex to any empty square on the board.

The initial positions of the game look like this:

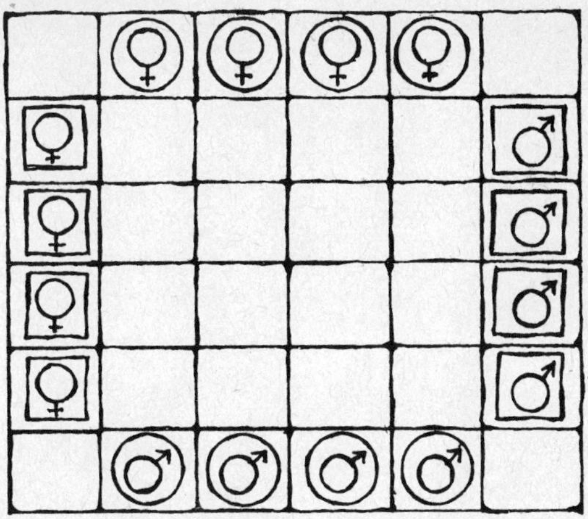

The object of the game is to prevent your opponent from being able to reproduce. One way to win, for example, would be to capture all of your opponent's males.

The theme of a game can sometimes carry with it strong social implications. In the game of Old Maid, for example, the loser gets caught holding the Old Maid card at the end. Recently a game called Robot has been marketed. The structure of the game is identical with that of Old Maid, but the old maid is replaced by a robot — the worst thing one can become is not an unmarried woman but a machine.

This modification of the game goes even further. Most Old Maid decks consist of twenty pairs of animals or people with silly names — e.g., Handy Andy or Silly Willy — and the Old Maid card. The Robot deck consists of twenty pairs of cards representing different occupations, such as judge, construction worker, mail carrier, scientist, etc. On each of the cards a man and a woman are shown doing the work. The faces are red, black, brown, yellow and white. Otherwise the game is identical with Old Maid.

I've observed young people playing Robot and commenting on the woman construction worker, the black scientist, the brown judge. Some of the youngsters have taken to using the word 'robot' to indicate unfeeling stupidity.

I can imagine other themes to be embodied in other games. For example, it is possible to create games centred about such charged subjects as:

— roaches and rats
— the game of school
— parents
— settling a new world
— the blue-eyed people vs. the green-eyed people
— getting rich by making other players poor.

2 Game boards

Most games manufactured for young children (four- to six-year-olds) are just decorated variations of an extremely simple game idea. The games are determined by the nature of the board and the fact that moves of the pieces are decided by chance devices such as spinners or dice. The board itself always has a starting point, an end point, and a series of steps in between. Therefore, basically its structure is:

though commercial game makers add a few twists and curves to make the board look more complex.

The simplest versions of these games consist of a race from start to finish. Usually some theme is attached to the game, so that simple race games are packaged under names such as Olympic Running, Speedway Thrills, Speedboat Racing, Air Derby, Finding the Treasure, Candyland, Gingerbread Man or The Pooh Game.

Some complex twists are often added by giving some squares special status. For example, some squares might say: go two moves forward; go back to start; take an extra turn; lose a turn; skip to the first blue square; etc.

Ray Nitta, a co-worker of mine at the Center for Open Learning and Teaching, made an interesting reading game out of a simple start-finish board. He put letters on each square of the board and then made a deck of cards containing the letters. Each player chose a different colour and moves were determined by a throw of one dice. The board looked something like this:

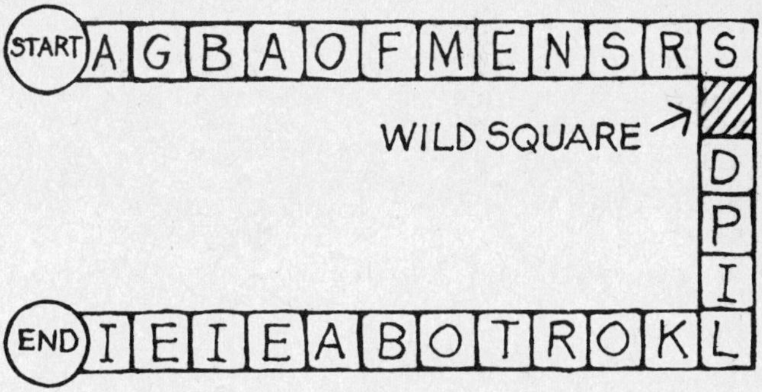

On the actual board each vowel appeared at least three times so that it was extremely likely that all the players would land on more than one vowel in a trip from start to finish. When a player lands on a particular square he or she gets a card with that letter on it. If anyone lands on the wild square they get a blank card which can stand for any letter later on in the game. Each player makes one trip from start to finish. After the last player is done the object of the game is for each player to make as many words as possible out of the letters he or she picked up from the board. Sometimes kids play in teams of two or three and pool their letters at the end. This makes it possible to spell out longer words and is a wonderful way to accustom young people to function collectively.

One scoring system we tried out with some second graders was:

— words of one, two, or three letters: each letter counts one point;

— words of four letters: five points;

— words of five letters: seven points;

— words of six or more letters: twice the number of letters in the word.

We decided upon this means of scoring so that there would be more incentive to make a few long words rather than many short ones.

Another friend, Ron Jones, discovered an educational toy

company that was going out of business. One day Ron showed up at my house with forty blank game boards. They all had the simple  structure, but there was nothing on the board to indicate how they should be used. I distributed a dozen of the boards to each of the kindergarten classes and a fourth-to-sixth grade multigraded class, and a half dozen in a junior high and suggested that students make up their own games. The kindergarten classes made up a dinosaur race, two races between cops and robbers to get a treasure placed at the finish, and a contest between superheroes. The children decorated the boards, made up special instructions on some of the squares (such as 'car crash, go back to start' or 'superman's special space — go five more if he lands here') and played the games for quite a while. One of the junior high school students made up a card deck of silly tasks (such as 'spin around until you fall down dizzy', 'kiss the person sitting to the left of you', 'pick a partner and pretend the two of you are going out together for the first time', etc.). People who landed on squares marked 'it's your turn to be silly now' had to follow the instructions on the card. The game didn't have other goals — reaching the finish first was identical to the fun of trying to get around the board with the fewest number of silly tasks to perform.

A high school guerrilla theatre group I knew expanded upon this game. We marked off a people-sized game board in a park. Half the troupe of twelve played the game at a time. The six players began at start and threw a dice, then moved along the board the proper number of squares. On each square was a card giving instructions for pantomime improvisions. Each time the game was played the cards were reshuffled so that one could never tell which improvisation was associated with which square on the board. The six members of the troupe not on the board had to watch the improvisations and guess what they were. There was no winning or losing in this game, which was more of a warm-up exercise for us than anything else. Some of the exercises we used were:

— you are getting up in the morning with a hangover
— you have just had your leg shot off in Vietnam
— you are at a boring cocktail party
— you are teaching a class of bored high school students

— you are trying to catch a taxi
— someone is trying to hunt you down and you are trying to escape.

A number of themes developed for this game were turned into scenarios for performances we elaborated on and many of them were themes we had used during writing class.

Next year I'll be teaching kindergarten and first grade. I plan to reproduce several blank game boards so that the children and I can make our own games. The boards will be eleven-inch by fourteen-inch versions of the following two forms:

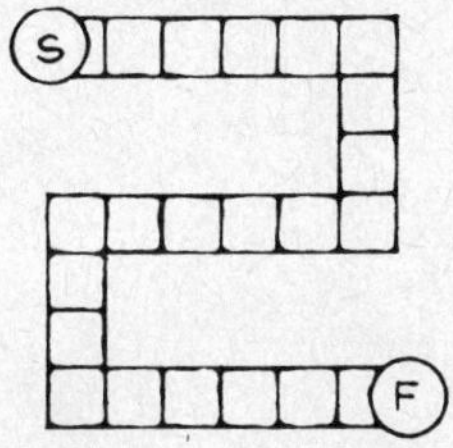

1.

2.

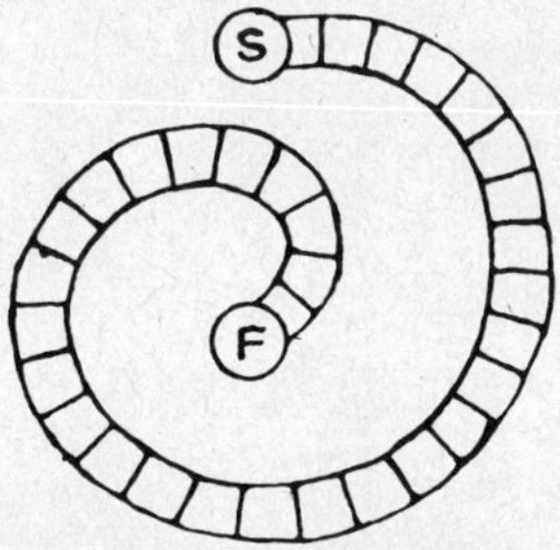

There are some other variations on this same idea that might make interesting games. For example, instead of one starting place there can be a series of different starting places with a common goal:

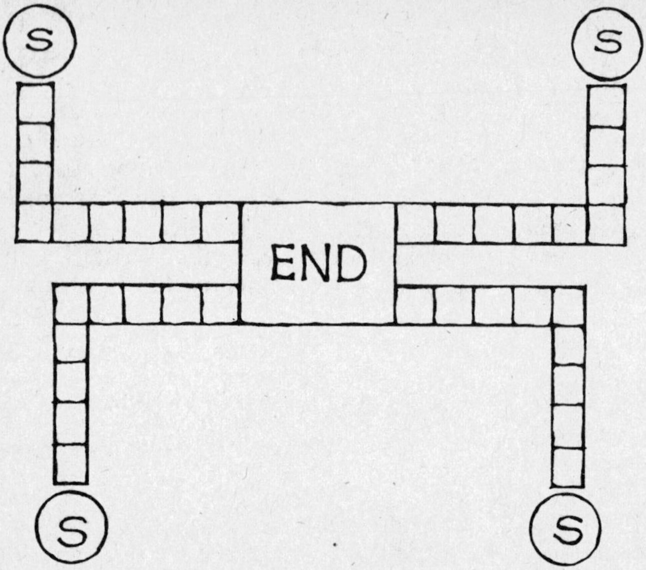

Each player could take a different route with different special squares. For example, in an animal game the board above can be elaborated on in the following way:

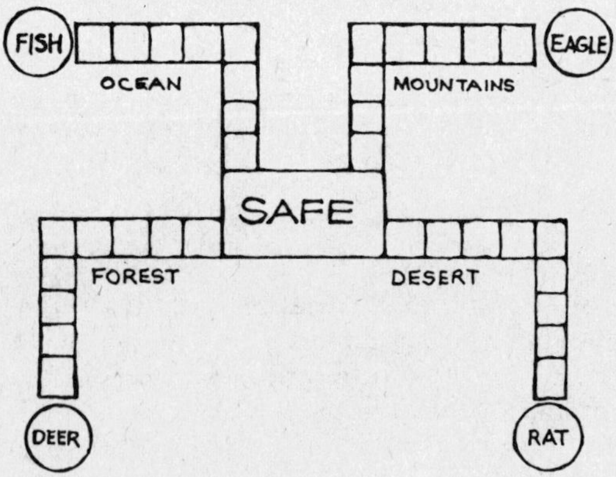

A plastic fish, eagle, deer and desert rat can be the pieces, and the privileged squares in each arm of the board can be made to represent something about the nature of the animals'

environment. For example, in the case of the desert rat:

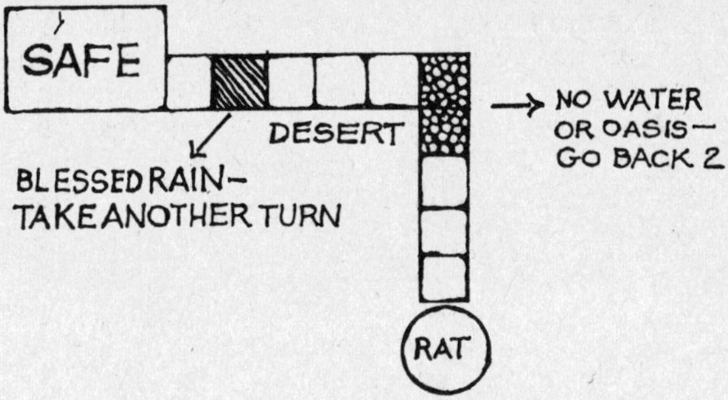

The game maker usually has to be careful, however, to make travel from each of the starting points equally easy or difficult. But there are times when it might be interesting to set up a game where one starting position made winning easy and to ask the students to figure out the best way to win the game. Such a board might look like this:

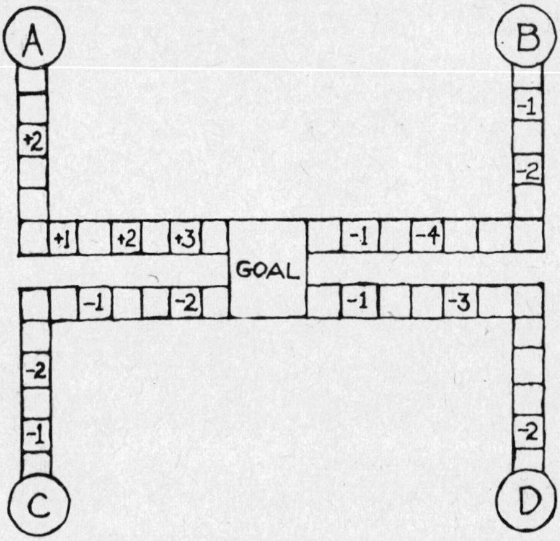

On this board (the pluses and minuses representing bonus moves ahead or backward), A is set up to have a much better chance of winning than B, C or D.* Many games like this one involve arithmetic operations and the more interesting to students than rote drill of mathematical facts. This game, for example, is an easy way to introduce the concept of negative numbers. For most youngsters the idea of moving backward is clear, while the idea of numbers less than zero can seem mystical unless it is concretely presented.

On the  board the players do not confront each other directly. Rather they move their pieces along the same path and, since moves are decided by dice or spinners, they have little control over the outcome of the game. There are many board games, the most familiar of which are checkers and chess, in which strategy rather than chance determines the outcome. In these games players usually face each other over a board and have a fixed number of pieces whose moves are determined by set rules. However, at most points in the game players have choices open to them — they decide how to move the pieces rather than depend upon chance devices. Most young children's games don't incorporate strategy in this way. Chess is usually not taught to four-, five- and six-year-olds, and I have even talked to teachers who believe that checkers is too difficult for children in kindergarten and first grade. The argument is usually that young children don't really understand how to make choices. Since most traditional schools discourage independence and reward obedience it's not surprising that games involving choice are considered beyond the students' capabilities.

I have experimented with many easy strategy games, playing around with variations on checkers and chess, and found that young people are much more engaged by strategy games than by chance games, which easily becoming boring.†

The checker board lends itself to many modifications and simplifications that can be used in many different ways in the classroom. The simplest interesting chess/chequer type board

*If the rules were that you had to have the exact number of moves to go into the goal to win, then it would not necessarily be easier for A to win in this game.

†Of course craps is a chance game, and lots of kids never get tired of playing it. However, the excitement of the game doesn't come simply from the casting of the dice. Rather it is the bluffing and strategy in betting that keeps the game exciting.

I've worked with is a three-by-three board:

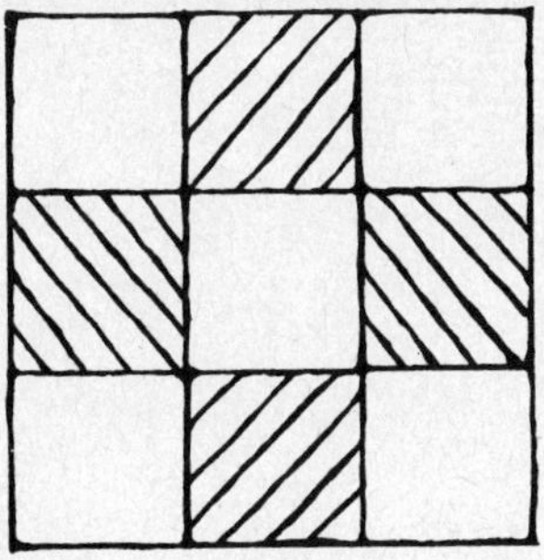

There are a number of games that can be played on such a simple board and, since most of the games are so short, a number of interesting mathematical and strategic questions can be studied using these games.

The most familiar game placed on a three-by-three board is noughts-and-crosses.

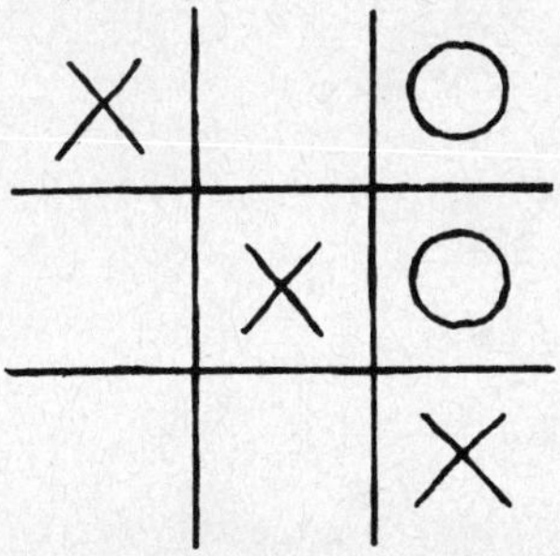

There is an interesting modification of noughts-and-crosses which turns it more clearly into a board game. Each player is allowed three tokens. (I usually use three red and three blue poker chips.) The board is three squares by three squares just as in noughts-and-crosses. The players take turns putting their pieces on the board, trying to achieve three in a row just as in noughts-and-crosses. However, if after all six pieces are placed on the board there is no winner, the players continue with the same six pieces and take turns moving them about the board trying to get three in a row. If the same position is repeated three times in a row a tie is called.

For example, after placing the six pieces the following position can result:

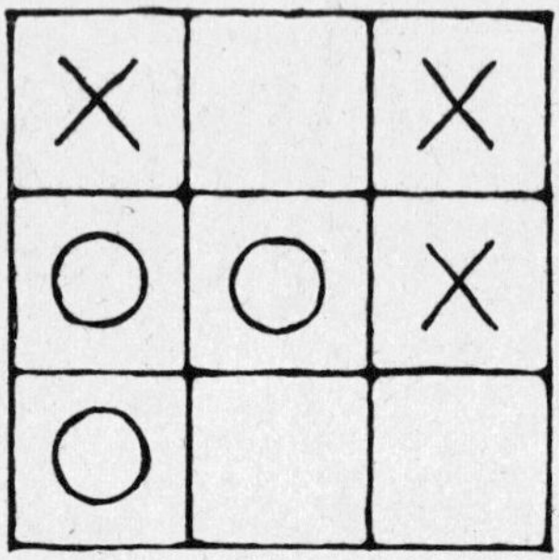

with X to move

In regular noughts-and-crosses, X would then go in the lower-right-hand corner and win.

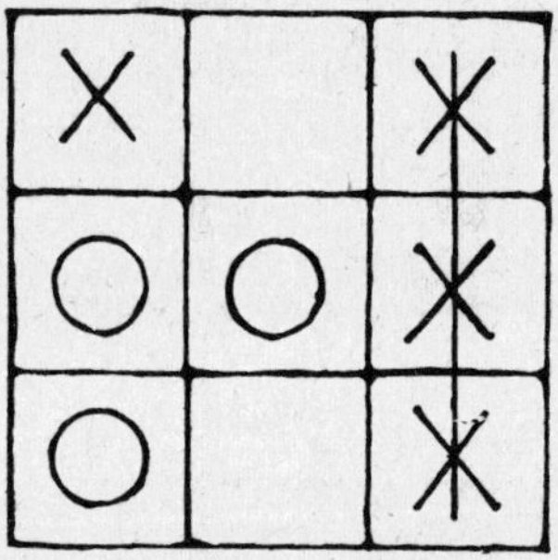

However, in the game I'm describing no additional pieces are added after the first six are put down. Using the six pieces on the board and adding the rule that a piece can move vertically, horizontally, or diagonally to any adjacent empty square, a new game arises. In the example, X could still win on the next move by moving to the top middle position:

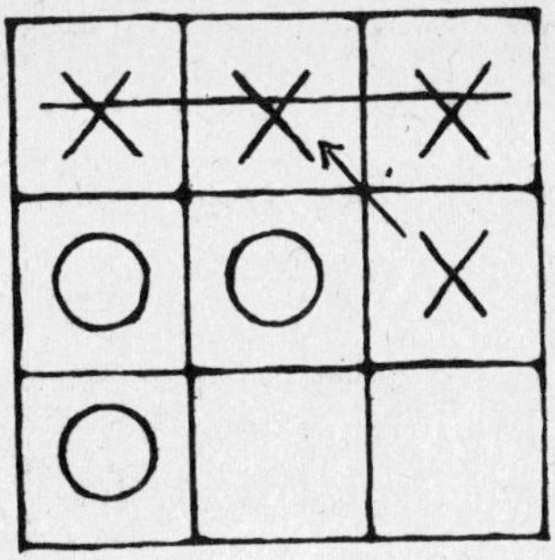

This is quite a simple board game that I've used with kids who find chequers a little too difficult. It provides a good introduction to strategy games using another game that most of the kids are already familiar with or can be taught easily.

There is another game played on the three-by-three board that I've used with elementary and secondary school students as well as with people in teacher training to introduce some basic mathematical concepts. The game uses six chess pawns, three black and three white, placed opposite each other on the board.

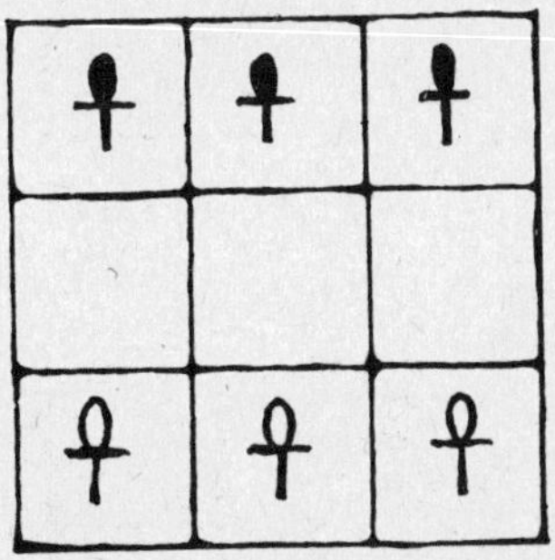

The pieces move as in chess — one square forward at a time, with capture on the diagonal only. Therefore, when two pawns are face to face neither can move unless they can capture diagonally. In the following diagram the black pawn can capture diagonally to the left or right but cannot capture straight ahead:

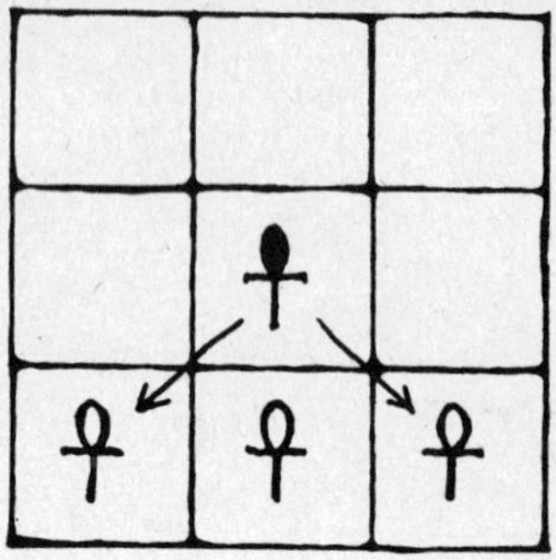

resulting in either of the following positions:*

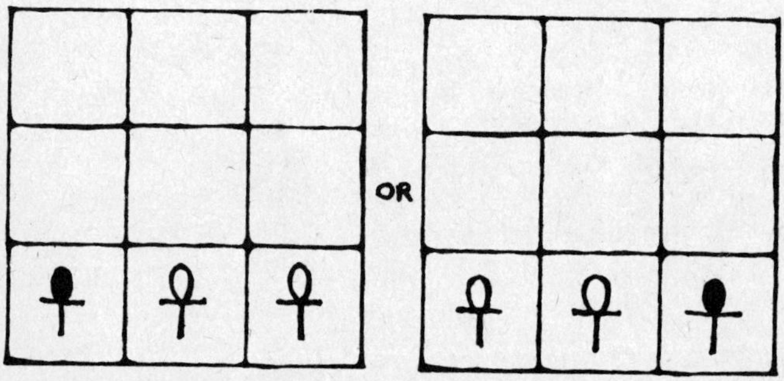

The winner of the game is either the first player to get a pawn to the opposite end of the board or the last person to move. Thus in the following situations with black to move a win is achieved.

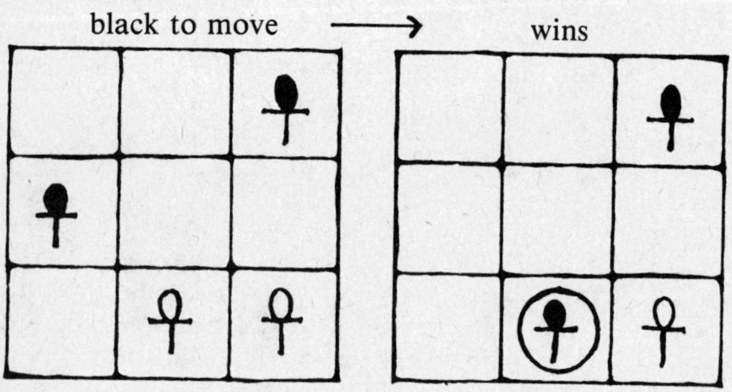

*As I describe this and subsequent games, I think it would help the reader to draw a board and, using poker chips or some other kind of token, play the games through instead of depending solely on the diagrams.

black to move ⟶ wins

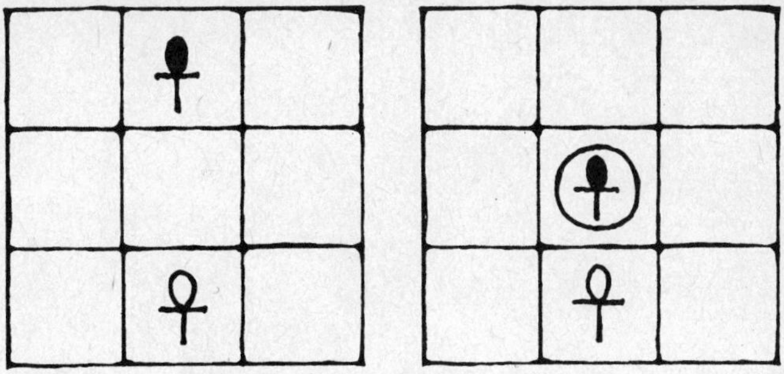

In the last example white cannot move since it is blocked by black's last move.

The game is very simple and usually lasts no more than three or four moves. I've found even adults challenged by the game and fascinated by its simplicity. Here is a sample game:

Game 1

opening position ⟶ first move white ⟶

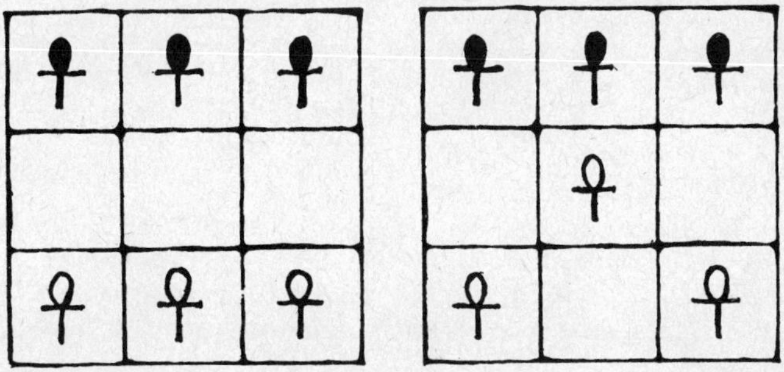

first move black ⟶ second move white ⟶

(takes white on centre square)

second move black

(takes white and wins)

Play the game a few times with a friend. Almost everyone who plays the game for a while begins to discover that there are strategies that can be used to maximize one's chances for winning.

After people have a chance to become familiar with the game, I pose the following questions:

1. How can you record a game you have won in order to allow someone else to replay the game and find out how you moved?

2. Can you figure out a way to play the game so you never lose?

3. Who has the advantage in the game, the player who moves first or the player who moves second?

There are a number of advantages in discussing simple games in this abstract manner. Many students become terrified of mathematics after a few years of school and avoid abstract symbols of any sort. The game which becomes the basis of abstract analysis is concrete, can be manipulated. It provides a physical and visual base upon which abstract and mathematical considerations can be built. The game is a mini-system which is very similar to a finite algebra. Moves are like mathematical operations; complete games are analogous to proofs; guesses about who will win or lose are similar to the creation of mathematical hypotheses. It is often possible to move from the analysis of simple games to the study of finite mathematics* and short-circuit the fears that people have built up about mathematics.

I gave one way of transcribing a game in the sample game above. The technique I used (of drawing a picture for each move) is time consuming and takes up a lot of space. Usually when a notation system to describe the game is developed (as in question 1) pictures are not used. I've found that kids as young as eight or nine as well as adults usually abstract from the picture method right away and begin to label the board and pieces.

Here are a few of the most common ways the board and pieces have been labelled during classes I've participated in:

The board

1	2	3
4	5	6
7	8	9

i.

1	8	7
2	9	6
3	4	5

ii.

(equivalent ways of numbering each square independently)

*A good text to use at this point is *Introduction to Finite Mathematics* by John Kemeny, J. Laurie Snell, and Gerald L. Thompson (Prentice-Hall, 2nd ed. 1966, $11.80).

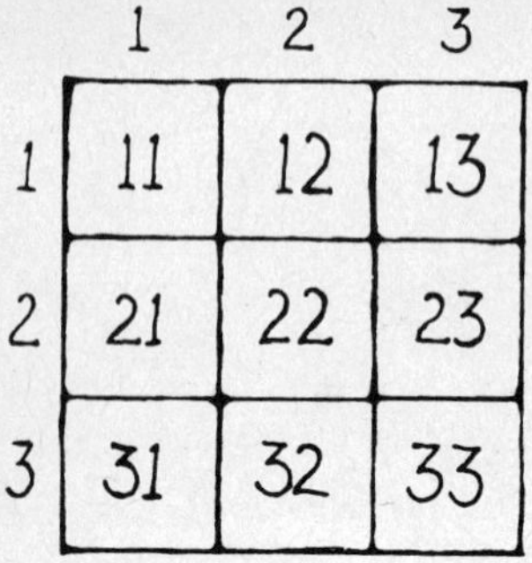

iii.

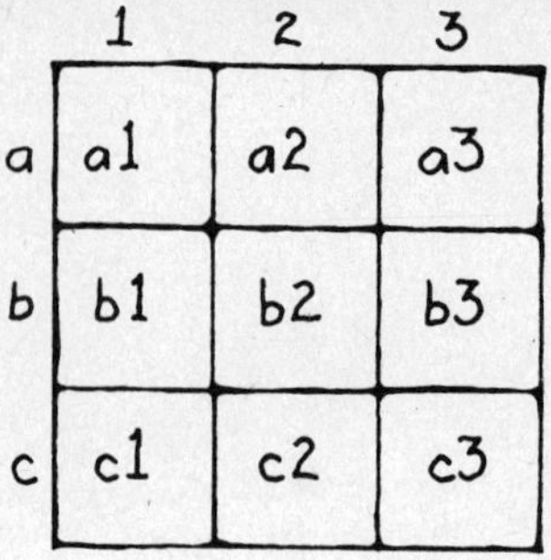

iv.

(equivalent ways of creating coordinate systems so that each square on the board is unique)

The pieces

Ways of labelling them

(The numbers 7, 8, 9 have been chosen because squares 7, 8, 9 reflect the starting position of white using board notation i.)

For the sake of clarity (another important mathematical notion) the third notation system for pieces is usually decided upon by the group, because it distinguishes between black and white pieces.

In deciding upon a game notation there are two other aspects of the game which must be indicated — the starting or *initial position* of the game, and a designation of which player is to move first (the *order of play*). The order of play is usually decided by some random device such as throwing a die or cutting a deck of cards. In some games, first moves are determined by custom, as in chess, where white always moves first and the decision is over who gets to play white.

Thus in the three-pawn game described above, if player 1 moves first, the initial position of the game can be represented as follows:

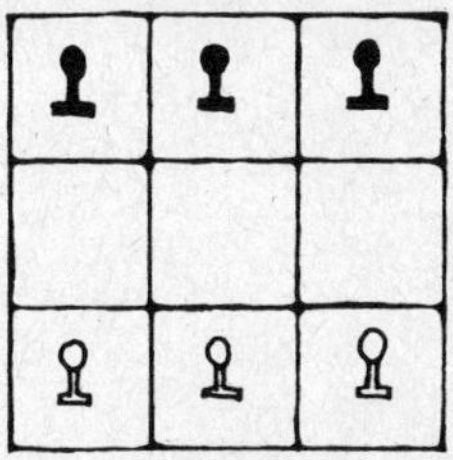

After labelling the pieces and the board it is usually discovered that several other ideas have to be represented in the notation system if the process of the game is to be described. The first notion is that of one piece moving from one square to another. The second notion is that of capturing, since two pieces cannot occupy the same square at the same time in this game. Finally, the point at which the game ends has to be indicated. Here are some of the symbols people invented:

moves:

a) Z → y — piece Z moves to square y
b) y1 → y2 — the piece on square y1 moves to square y2

captures:

c) Z1 X Z2 — piece Z1 captures piece Z2
d) Z1 X y2 — piece Z1 captures the piece on square y2
e) Z → y (X) — the piece Z moves to square y and captures the piece already there (this combines two notions, that of moving and that of capture)

wins:

f) Z → y ! — piece Z moves to square y and wins
g) Z → y (X)! — piece Z moves to square Y, takes piece and wins
h) other symbols used to designate winning are:

↑ ✱ ☺

Using notation a) for moves, e) for captures, g) for wins and i) for the board, the following is a transcription of the sample game:

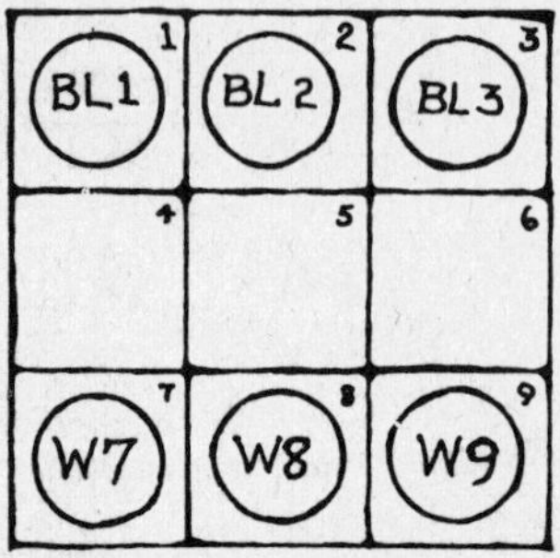

initial position of sample game with notation

move number	player white	player black
1	W8 → 5	BL1 → 5 (X)
2	W7 → 4	BL1 → 9 (X)!

Try to play the game from this notation scheme and try to transcribe a few games you have played in some notations you find clear. There is no best notation system, though some ways of representing the game are shorter and clearer than others. It is the same with all kinds of symbol systems which are invented to represent some activity beyond themselves. Some choices of symbols are arbitrary but the whole has to be governed by the idea that other people using the system can understand and often reproduce the events the notation represents.

The invention of a notation scheme for oneself seems to make it easier to understand other traditional systems of notation. I introduced standard chess and chequers notations which even seven- and eight-year-olds were able to figure out and record for themselves after having played with the simple game above. They also learned that there are at least two different major notation systems used for chess, the algebraic notation system and the descriptive notation system, each with its own advantages and disadvantages.

The algebraic notation system is simpler, more mathematical. The descriptive is more imagistic. Different players prefer to use different notations though there is a drive for a single notation system to facilitate communication among chess players.*

Games are not the only form of activity that can be recorded by abstract notations. Music immediately comes to mind, as well as dance, speech, counting and other mathematical operations.

Abstract notations provide manuals that enable people to reproduce the thoughts or actions of others. Musical notation enables people to re-create with some fidelity melodies they have never encountered before, just as the written language enables people to read words they have never heard spoken. One of the main goals of education is to give young people access to as much experience as possible. To master an abstract notation system opens up a world of otherwise inaccessible experience. Of course written language is the abstract notation that most obsesses our schools. But mathematical, musical, scientific or dance notations could get serious attention as well. It is important that young people feel that many different notations are accessible to them and that there are not certain aspects of the world that are beyond their capacities. I feel that to achieve this goal young people have to understand that notations are developed by people trying to record and convey that they have experienced and understood. Over the past few years I have experimented with having young people develop their own systems to record sound and movement as well as learn traditional notation systems.

Another co-worker at the Center for Open Learning and Teaching, John Rosenbaum, has developed several ingenious musical instruments which can be made and tuned by young children (my three-year-old son makes his own instruments).

One of the instruments is a do-it-yourself xylophone. The instrument consists of a simple frame constructed of wood in the form of a truncated triangle with an empty centre (these frames are anywhere from twelve inches to three feet long).

*For a good account of chess notation, see *Chess in a Nutshell* by Fred Reinfeld (Pocket Books, 1969, $1.25).

bottom view:

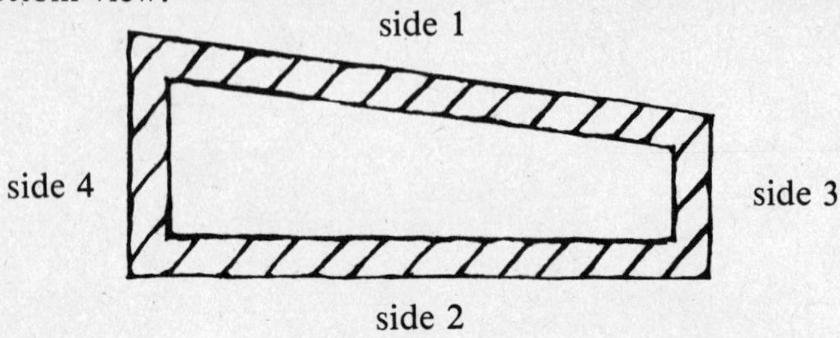

side view:

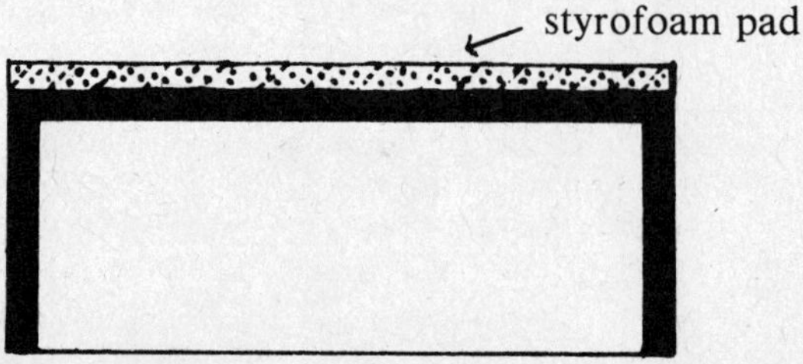

On the long sides (sides 1 and 2) a styrofoam strip the same width as the wood (from half to one inch wide) is glued down. That is the xylophone frame. Along with the frame comes a box of metal pipes, strips of wood and plastic, and a few strikers made out of superballs stuck on to chopsticks:

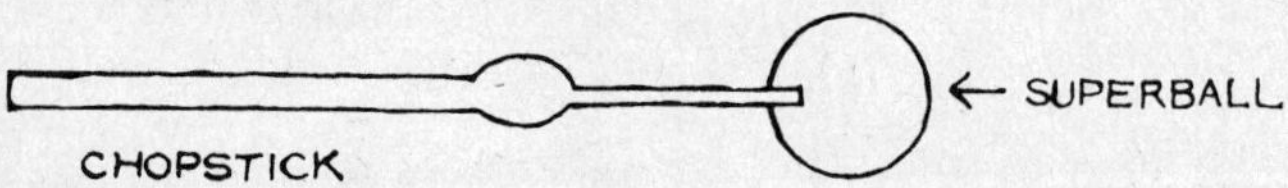

Any number of playable xylophones can be created using the kit. The sounds the instrument produces depend upon which strips of wood or metal are chosen to be the keys as well as how they are placed across the frame. For example, one night my daughter made a xylophone consisting of three pieces of walnut and metal ruler arranged over the frame as follows:

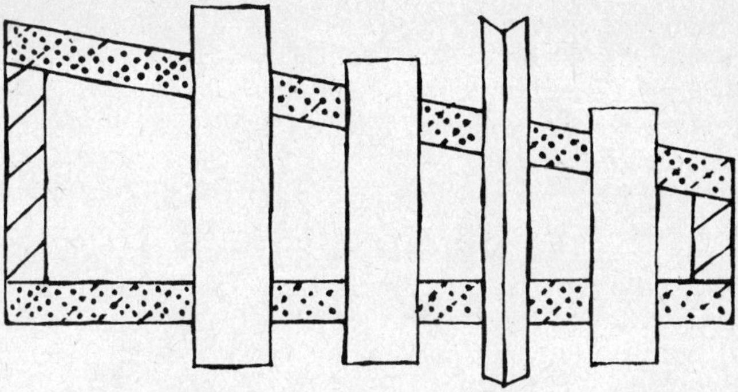

With this simple instrument she created a number of tunes and taught them to her sister. Then she asked me if it was possible to write down the tunes so she could play them another day. We sat down to figure out a notation system which she could use to preserve her composition. First we numbered the pieces of wood and metal on the frame 1, 2, 3, 4 and wrote these numbers on the pieces themselves so she could reconstruct the instrument in the future. Then she played the tune and wrote down the appropriate numbers: 1, 2, 4, 4, 2, 1, 3, 2, 4, 4. That was the extent of the notation she needed for her purposes. The length each note was held, the number of beats per bar and other aspects of traditional music notation were irrelevant to her at that point.

I have seen many more complicated notations created for improvised instruments by students at Other Ways, an alternative high school in Berkeley I worked at for three years. One class I taught was called Math Through Games. I taught the students the three-pawn game described above, asked them to create a notation system, gave them a range of possible game pieces (such as dice, spinners, boards, little animals, cars, soldiers, etc.), asked them to make up new games and create notations for the games they created. We did this a few times and I began to move on to mathematical notations and systems. One of the classes was interrupted by the students who showed us a notation system they were working on for guerrilla theatre.

The students in guerrilla theatre were creating their own

musical instruments — using cardboard flutes, glasses and jars, as well as rattles and drums improvised from found materials. Occasionally a session that began as pure improvisation ended up with a composition everyone felt was good enough to record or at least play another time. The students naturally and with no need of prompting from me transferred the experiments in making game notations to the realm of sound. They labelled the instruments (e.g., tube 1, tube 2, tube 3, jar 1, jar 2, jar 3, etc.) and developed a grid with even beats (but no bars of set rhythm) where each instrument had its own line. I remember it looked something like this:

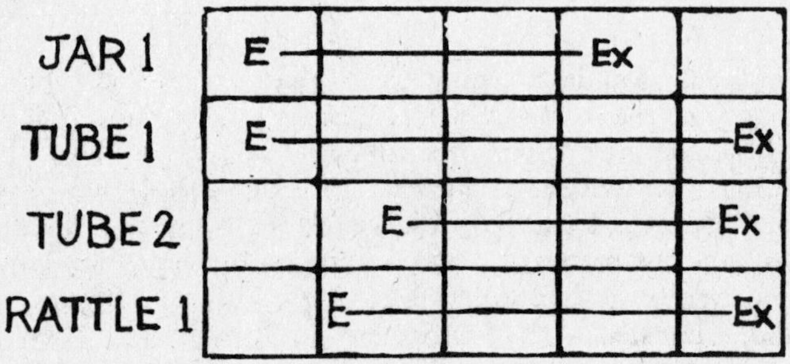

symbols:

enter and begin playing: E____________________
end playing: ____________________ Ex
continue to play: ____________________

This notation, which was later made much more complex and precise, leaves an incredible amount of room for improvisation since though it prescribes the order of entry of instruments it doesn't say anything about melody or rhythm. The above simple scenario can be realized in performance in many different ways.

Creating a new notation system doesn't preclude learning the standard notations. Rather it leads naturally to it. We considered the standard system of Western musical notation after having created our own system. There were obvious advantages to the traditional notations which indicated melody, rhythm, phrasing, etc. Our work with our own musical notation was excellent preparation for students who wanted to read music and play instruments.

The students also discovered some of the limitations of the twelve-tone scale which is central to Western notation systems. Some of the sounds the students made with their improvised instruments had no equivalents in the twelve-tone scale. There were quarter tones and eighth tones, sounds and rattles and wooden tubes that couldn't be fitted as easily into the Western scale as could piano or guitar chords.

One student who was studying Chinese, Indian and Arabic music described different tonal systems requiring different notations. Just as in the study of games, we saw how systems can change or be modified, and how they vary according to function.

A number of the same students tried to create a notation system for dance movement and found it extremely difficult to represent all the aspects of movement adequately. Eventually they settled on one notation which was like a map and represented a dancer's movement through a particular space and another that indicated body movements while the dancer remained in the same place. The form notation began with an actual floor plan of the space used by the dance performance and became a route map of the dancer's tour through the space:

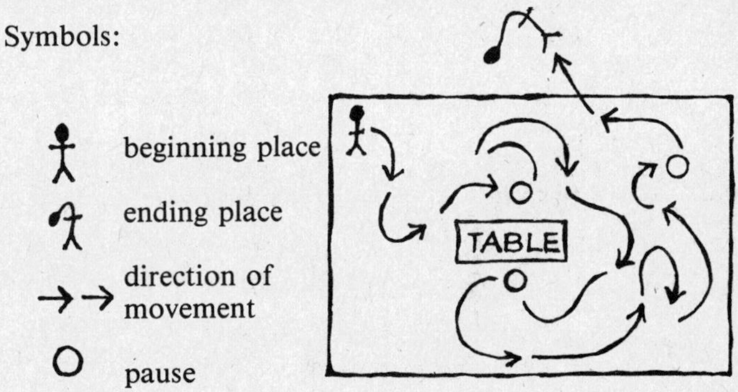

I seem to have come a long way from talking about games — but this is what always happens. Things that people learn through making and playing games pop up everywhere.

Game situations provide nonthreatening models for experimentations as well as for the mastery of skills involved in strategic thinking, such as the ability to entertain different possibilities, to see the implications of different choices, and to devise comprehensive plans of action and concrete ways to carry them out.

Because playing a game involves practice, playing a game over and over until it is mastered or becomes boring, the skills acquired almost become habitual. When people are encouraged to value what they learn in informal ways, much of the dread of mastering new and unfamiliar material disappears. I believe that most people are not aware of how much they know and how possible it is to control their own education. Of course mastering games or the ability to think strategically does not guarantee that these skills will be used humanely. We know that cruel and inhuman uses of game strategy have been made by the Pentagon and even by city planners. No skill can be considered apart from the moral situation in which it is used.

* * * *

At this point I want to return to the simple pawn game though from a different perspective. After having developed a notation system, one can explore the question whether a strategy exists for winning the game all the time. Of course this exploration presupposes that people have played the game often and have begun to discover inductively certain moves to avoid and others to make whenever possible. Let's look at the board with the initial position of the pieces using one of the board notation systems (iv.) illustrated above:

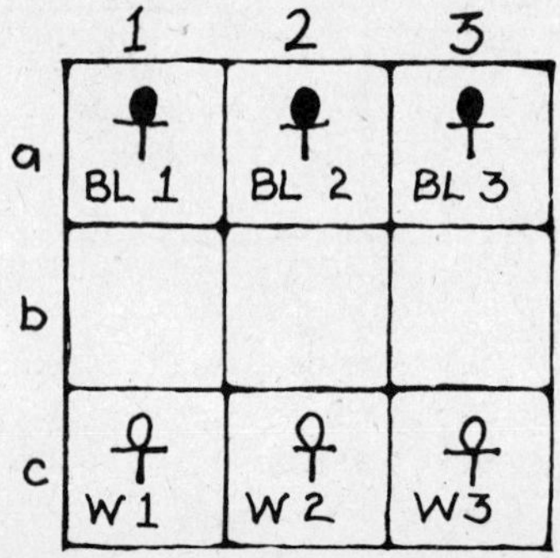

With white going first, there are three possible first moves:* W1 → b1; or W2 → b2; or W3 → b3.

Many people notice that it is not really necessary to consider all three moves since the W1 and W3 moves are mirror images of each other and any argument that can be made about moves of W1 and responses to that move can also be made about W3. Discovering this (as I have seen a number of fourth graders do) helps people to understand one of the important principles of mathematical reasoning — the reduction of a seemingly very complex problem to a series of simpler problems. Such simplifications can even lead to mathematical discoveries. When the great mathematician Gauss was a schoolboy his teacher told the class to add up all the numbers from 1 to 20. Almost immediately and without any apparent calculation Gauss wrote down an answer. The teacher seized the paper and called the class to attention, intending to humiliate the young Gauss. To the teacher's astonishment the answer was right and it was not merely a guess. Most students dealt with the problem by direct adding: 1 + 2 + 3 ... + 18 + 19 + 20. Gauss saw something quite simple that reduced tedious adding to three simple operations. He noticed that

$$\begin{aligned} 1 + 20 &= 21 \\ 2 + 19 &= 21 \\ 3 + 18 &= 21 \\ 4 + 17 &= 21 \end{aligned}$$

and so on. By pairing the numbers from one to twenty:

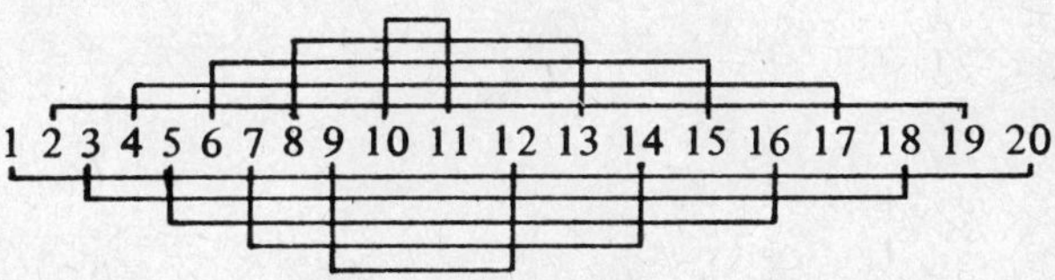

he got a series of 21s. Since there are ten pairs in the first twenty numbers, the sum of all the numbers is 10 x 21 = 210.

*It is easy to follow this argument if you set up a board with the same labels, use chips or markers, and play through what is described.

Three simple operations gave him the answer: (1) 20 + 1 = 21; (2) 20 ÷ 2 = 10; (3) 21 x 10 = 210. The same holds true for the sum from 1 to any even number. For example the sum of all the numbers from 1 to 100 is: (1) 100 + 1 = 101; (2) 100 ÷ 2 = 50; (3) 101 x 50 = 5050.

* * * *

Let's return to the analysis of the pawn game and the mirror moves. If you don't see how moves of W1 and W3 mirror each other and therefore are two cases of a single way of playing the game, play some additional games using two boards instead of one, making moves that reflect each other.

Sample mirror games
(X means takes; ! means wins)

Move	Player W	Player BL	Player W	Player BL
1	W1→b1	BL2→b1(X)	W3→b3	BL2→b3(X)
2	W2→b2	BL2→c1(!)	W2→b2	BL2→c3(!)

In order to figure out a winning strategy, following the previous argument, there are only two first moves that need be considered: W1 → b1; W2 → b2.

Take the first case: W1 → b1

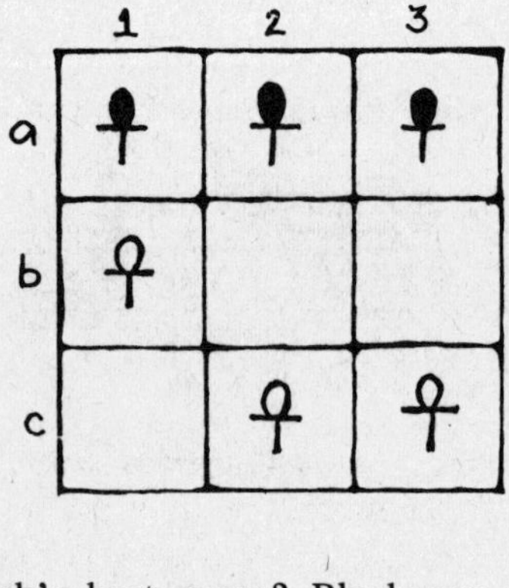

What is black's best move? Black can only make three replies:

	1	BL2→b1 (X)
or	2	BL2→b2
or	3	BL3→b3

Suppose black chooses the third possibility and moves BL3 → b3. Then white moves W1 → a2 (X) and wins.

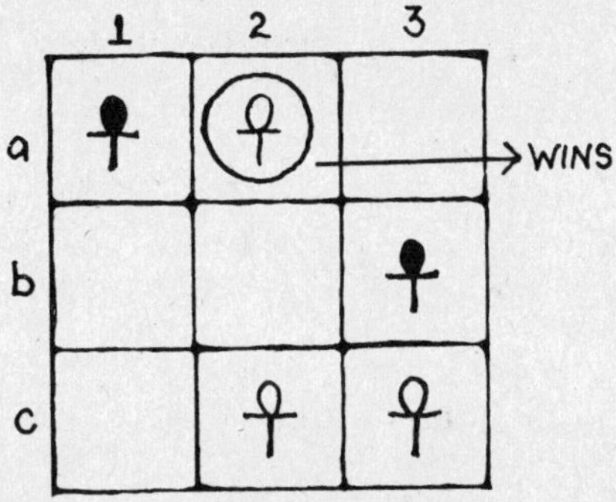

Clearly black must choose another move. How about: BL2 → b1 (X).

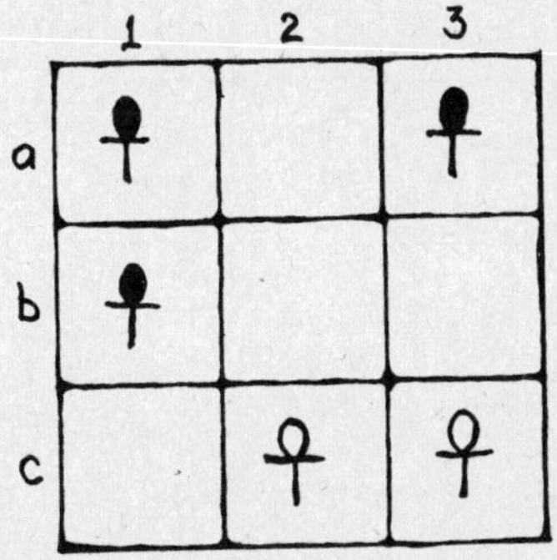

Now white has to take BL2 (which is on b1) or black will move to c1 and win the game (or take W2 and move to c2 and also win).

After white takes BL2 the board looks like this:

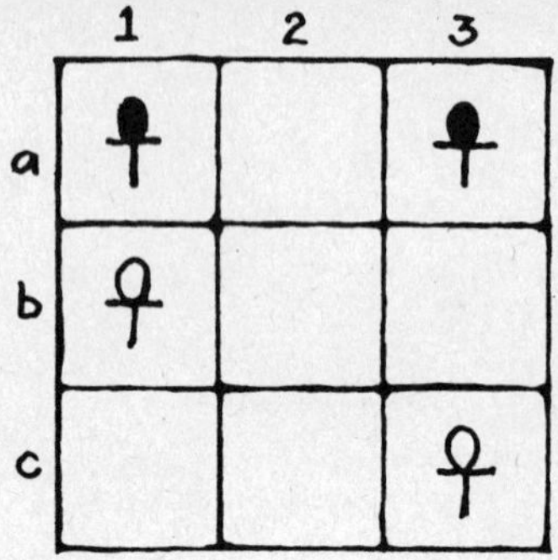

Now black has only one move BL3 → b3, which is a winning move since white is then pinned down and cannot make another move.

There is only one other possibility for black's first move in reply to white's first move: BL2 → b2. The board would then look like this:

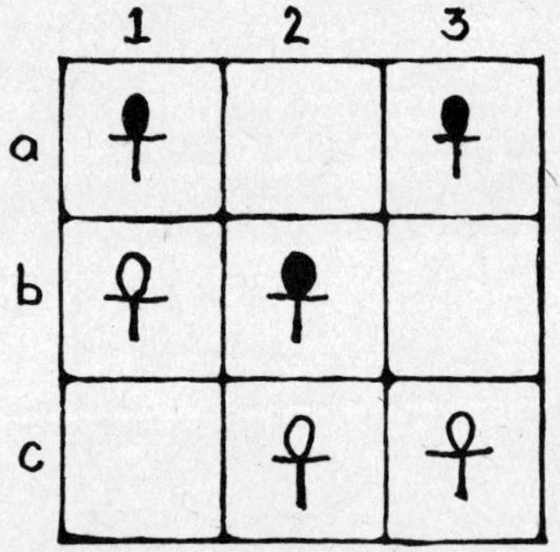

White can then move W2 → b3 and, by pinning down black, win — so this is a dumb move for black to make.

To summarize the argument, if white makes a first move W1 → b1, there are three possible responses yielding the following results:

W1 → b1 — BL2 → b1 (X) wins
W1 → b1 — BL2 → b2 loses
W1 → b1 — BL3 → b3 loses

Black has three responses to white's move W1 → b1, two of which provide losses and one of which guarantees a win, if black (the second player to move) plays intelligently. So far there are good reasons to want to be the second to move in this game.

Try to analyse the consequences of white's first move being W2 → b2. Is there any way for white to win if black plays intelligently? What moves provide guaranteed wins? How can you prove your strategy must work?

An analysis of the rest of this game is provided in the appendix. Try it yourself or with your pupils. It is a simple way to get involved in mathematical and strategic thinking, one which I have found is often extremely difficult for adults but easy for young people who don't think of the exercise as academic maths but as a way of acquiring a new trick.

* * * *

The three-by-three board I've been discussing is very simple, though the analysis of the game might initially seem difficult to people unfamiliar with abstract analysis. However, as soon as the board is made a bit more complex the analysis of games moves beyond most people who are not professional mathematicians or game freaks. For example, play the pawn game described above on a four-by-four board and attempt some analysis of a winning strategy:

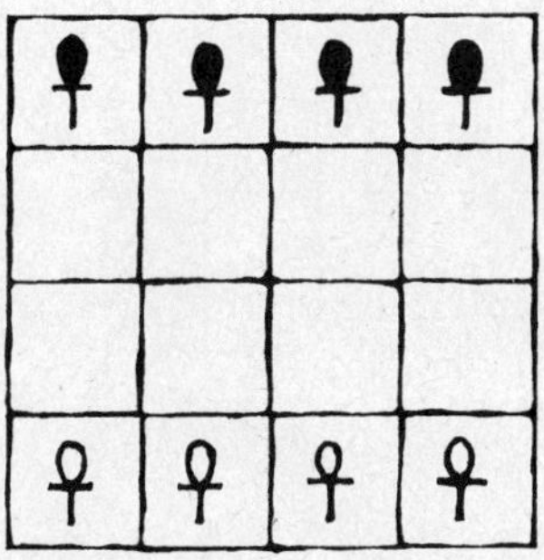

As an aid to teaching games and to creating situations where analysis of game strategies is not beyond most secondary school students, I've experimented with simplifications of games. This can often be achieved by reducing the number of squares on the board and the number of pieces while retaining all the rules governing moves and winning. Here are some samples of minimal game boards with the pieces placed in the starting position of the game:

1. Minimal chequers, played just like chequers but simpler and quicker to play, often more fun for very young children than a full game.

four-by-four board

two pieces on each side

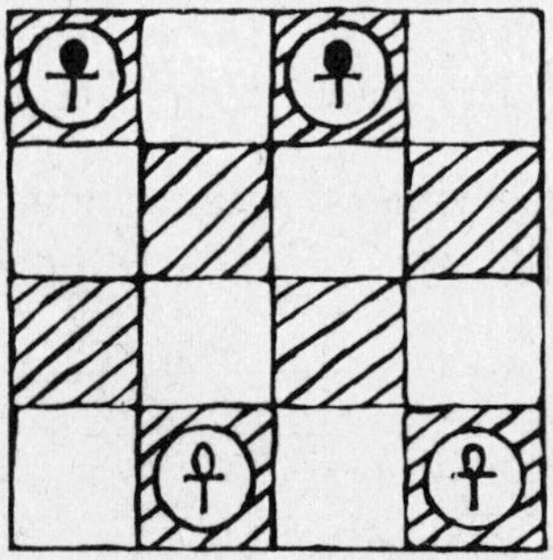

2. Minimal chess: the smallest board and game using all the pieces and rules of chess, not a particularly good way to introduce the game of chess (the whole board seems better) but a good way to get people who know the moves and rules to think strategically.

five-by-five board

pieces: each side has five pawns, and one each of king, queen, bishop, knight, and rook.

3. Minimal Go: Go is played on the intersection points of the board and not on the squares. As an introductory game (especially effective with children over five), I've found a

six-by-six board with handicap spots placed in the middle (as illustrated below) a fine and simple introduction to this sophisticated game. (For books on how to play, see the bibliography at the end of this chapter.)

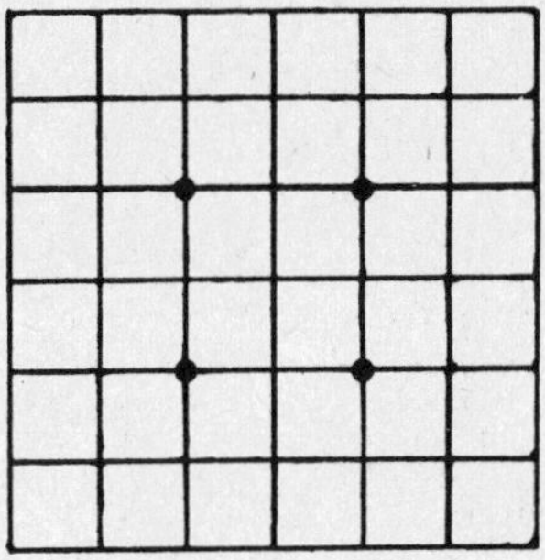

4. An interesting and simple mathematical game is packaged commercially under many different names. I have seen 99-cent versions as well as ones at $12.00. Essentially the game consists of a pair of dice and nine number cards (with numbers printed on only one side) laid out in a row.

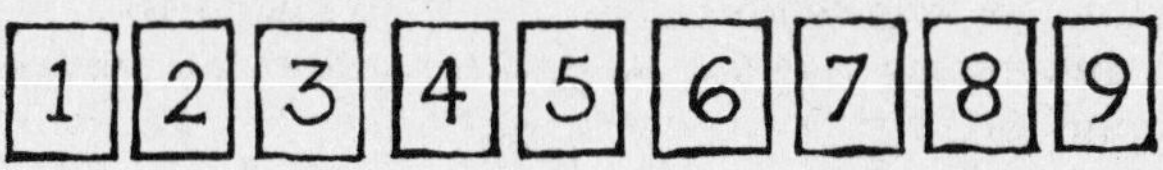

The game can be played by any number of people. On each turn a player casts the dice over and over, each time turning over cards that add up to the number cast until it is not possible to turn over any more cards. The sum of the remaining numbers is the player's score. Lowest score wins.

Perhaps an illustration of one player's turn will make this clear.

Player 1 casts the dice and gets a 9 (6 on one dice, 3 on the other).

He or she can turn over any combination of cards that add up to 9. This 3 and 6; 4 and 5; 1, 3 and 5; etc.

Suppose 1 and 8 are chosen. Then the board (which in this case consists of the cards laid out on a flat surface) looks like this:

The *same player* casts again. Suppose an 11 comes up and the player chooses to turn over 7 and 4. Then the board changes to:

Then he or she casts again and comes up with another 11 and this time chooses 5 and 6. The board then looks like this:

and the same player casts once more. This time let's suppose 6 comes up on the dice. No combination of 2, 3 or 9 — the only numbers left — can make up 6 if the game is restricted to adding.* Therefore that player goes out with a score of 2 + 3 + 9 = 14, since those cards are left showing.

This is an extremely simple game, cheap to make and easy to play. Most kids from the second grade on are fascinated with it and for some children just learning to add and subtract it is an interesting way for them to practise those skills with small numbers.†

I was showing this game to a group of teachers. Several of the teachers worked with four- and five-year-olds and complained that the game was too complex for their students. After some thought, Maria Rosa Gruenwalt, one of the teachers at the Early Learning Center in Berkeley, suggested the following simplification which she later tried with her students and found to work well.

Instead of nine cards six were used

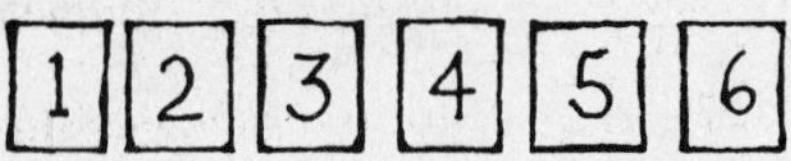

*Of course the rules can be changed to include subtraction as well, as was suggested by a fourth grader. Then 6 could be made by 9 and 3.

†There is no reason why the game can't be made more complex by using numbers from 1 to 20 and including multiplication and division, though I've never tried this.

and instead of two dice only one was used. Otherwise the rules were the same. For students who couldn't do the adding themselves, bead-counting frames were provided so they could figure out the answers by counting on the beads. This illustrates a very important point. Simplifying games, making games more complex, changing the board or the pieces or the rules are good exercises in freeing oneself from the passivity induced by depending upon preprogrammed materials and objects. It gets people accustomed to building for themselves.

5. Probably the most familiar game board to most of the peoples in the world is the Wari or Mancala board which in its most common form has two rows of six cups bounded by two larger cups:

The game of Wari originated somewhere in central Africa, spread throughout the African continent, and then under the name of mancala was carried by Moslems through Asia. It is played from the southern tip of the African continent to Hawaii. There is even speculation that representations of Wari boards can be found on some Mayan, Aztec and other pre-Western American documents. It was also brought to the West Indies and the Americas during the seventeenth century by Africans who were enslaved. I first discovered the game under the commercial name Pitfall. The board was manufactured in West Germany and there was no indication whatever that the game was African in origin and widely played.

Wari is a strategy game for two people, like chess and Go. There are dozens of variants of the game, though some principles of play are common to all. The game is played with stones (or shells or coins or nuts) which are placed in the two rows of six cups. There are three or four or five pieces placed in each small cup depending upon the complexity of the game. For simplicity I'll describe a three-piece game.*

*If you wish to follow the game on a board of your own, an empty egg carton and two paper cups provide a usable board, with beans for pieces.

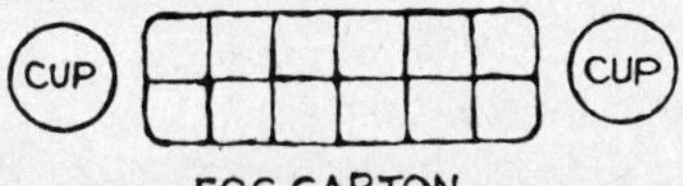

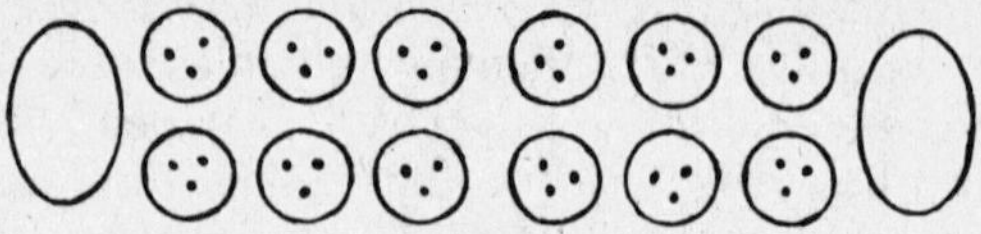

The African versions of Wari move from left to right around the board (though the Philippine version moves from right to left) and will be used here. The players sit opposite each other and the small cups in front of them are the places they move from. The large cup to each player's right is his home base.

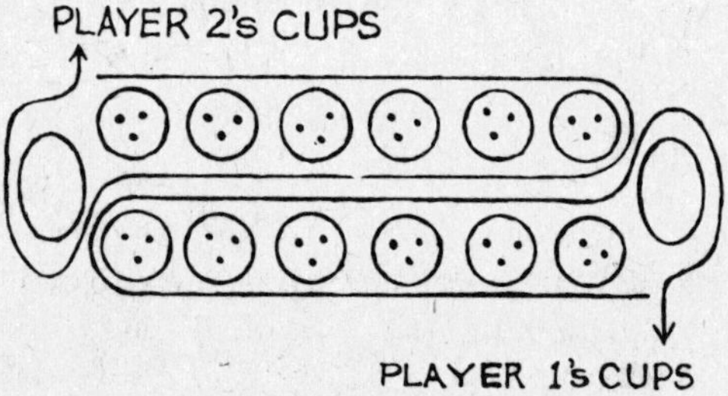

The simplest version of the game, and the one I usually teach first, has only two rules.*

First, a player moves by taking all the pieces out of any one of his six small cups and distributing them, a piece in each cup, in sequence around to the right. If a piece lands in the player's home cup it is his point. If after putting a piece in home base there are still pieces left, the player puts the pieces in his opponent's cups in the same way.

Sample move:
1. starting position

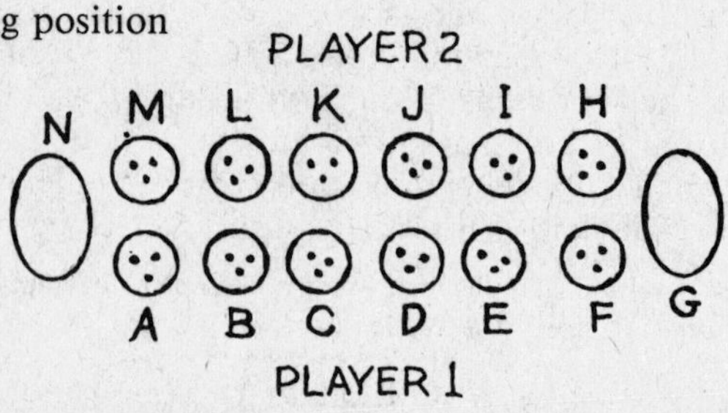

*For more complex rules and an illustrated history of the game, consult the pamphlet on Mancala listed in the bibliography at the end of the chapter.

2. after the move

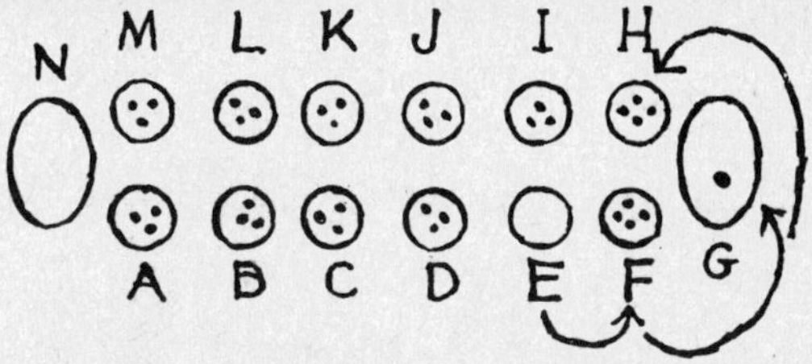

The second rule is that if at the end of a move the last piece lands in an empty cup of one's own, then all the opponent's pieces in the cup opposite are captured and put in one's own home base.

Sample capture move (at some later point in the game):

1. player 1 to move from cup B

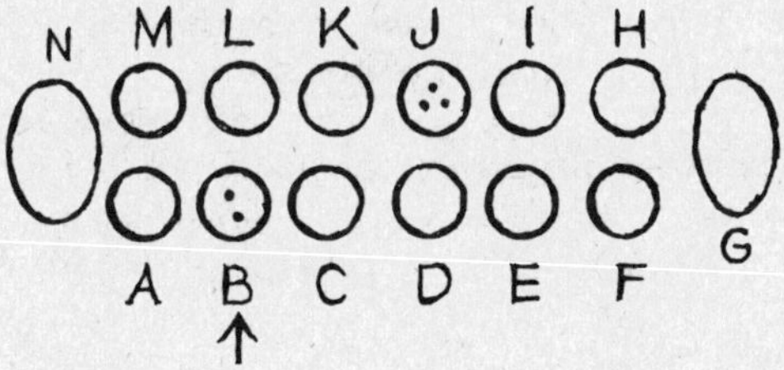

2. result of move

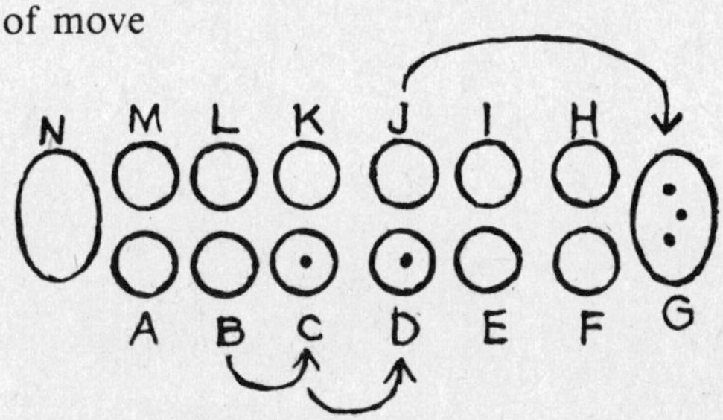

The last stone from cup B ends up in cup D, which is empty, so player 1 wins what is in player 2's cup opposite (J) and puts those pieces in his or her home base (G).

The game ends when one player has no more pieces on his or her side. When this happens the score is tallied in the following way:

1. Score for person to go out = number of pieces in his or her home base.

2. Score for person left with pieces = number of pieces in home base minus number of pieces left on board.

Sample scoring with player 2 going out first:*

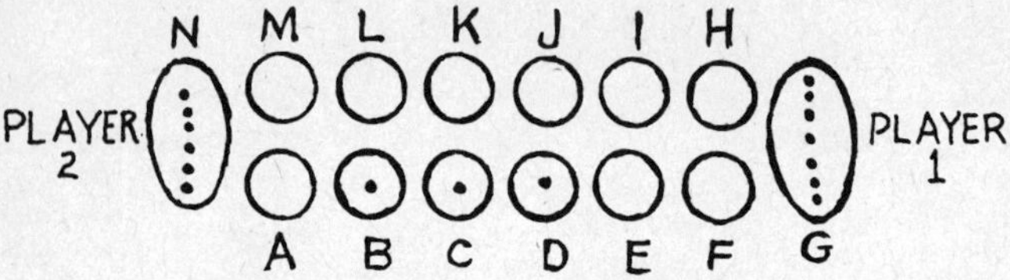

player 2: 6 player 1: 7 — 3 = 4

There are two things to note about Wari for players unfamiliar with the game:

First, the pieces all look alike, since they change in status according to which side of the board they are on.

Second, the player who goes out first does not necessarily win and therefore simply going out is not the goal of the game. Here for example is a sitution in which player 2 goes out first and loses:

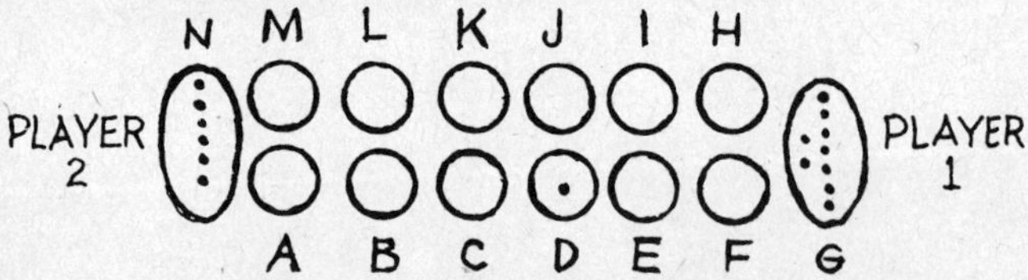

score

player 2: 6 player 1: 9 — 1 = 8

With a lot of practice Wari becomes a quick, exciting and complex game to play. The number of variations on the game and its use throughout the world make interesting social and cultural history. There are some teachers in a fourth- to sixth-grade classroom in Berkeley who have designed a whole social studies curriculum centring around studying the game, and have even named their class (which is an alternative

*The illustration of scoring is simplified here by not showing all of the thirty-six pieces which would exist in an actual three-piece game.

sub-school in a traditional public elementary school) Wari School.

I have taught Wari to five- and six-year-olds and found it difficult. The game is too long, the strategy is hard to grasp and the experience of trying to learn the game is too frustrating for the children to enjoy the game itself. Therefore I have simplified it and found the younger children have no problem playing the game on a somewhat smaller board (four cups on each side) with two pieces in each cup.

Initial position:

It is important to realize that other simplifications can be made. It is just a question of playing around until you discover what is most effective with your students. Sometimes it helps to share the process of simplifying the game with the students — show them the complex game, explain how it can be changed, and engage them in the dual process of learning the simple version and then generalizing what they have learned to assist them in mastering the more complex version. This skill of generalizing from simple models can turn out to be useful elsewhere, such as in maths and science.

I have used Wari in a very different situation. The schools in Berkeley were desegregated seven years ago. However, most of the white children come from upper-middle-class families, while many of the black children come from poor families. In addition, despite efforts by some people within the district, the curriculum and tone of most of the schools reflect the values of the American middle class. This puts the poor children at a disadvantage. Within the district, the average scores of the white children are above the national norms and those of the black children below. Last year I worked in a game centre at Hillside School, one of the public primary schools. My students were black and white in grades one to three. I started with chess and quickly ran into trouble. A number of the white children already knew the game, while none of the black children had any experience with it. If I stuck with chess the black children would be put in the position of learning what the white children had already learned at home. After a day I abandoned chess and switched to Wari. None of the children had played Wari before and so everyone started out as equals. Some students played better than others but economic class didn't influence skill. Learning something new to everyone, the students could work together much more smoothly.

* * * *

There are a number of other kinds of game boards worth discussing. The simplest game that uses a board and one very popular with young children (three to six) is lotto, or the matching board game. There are many versions of lotto, all of which consist essentially of a number of boards, each with a series of pictures on it, and a deck of cards of the same pictures. The object of most of these games (though the specific rules differ from game to game) is to match the cards to the right pictures on the boards, thereby filling up the boards.

Lotto games are used in most primary schools to help develop visual perception and awareness of colours, shapes and sizes, differences and similarities. They are also used to give young children practice in sorting objects into different categories, such as animal, plant, flies, swims, four-legged, man-made, mineral, etc. Most programmes intended to prepare the way for reading make use of the crude variety of commerically-produced lotto games that can be bought in any toy store.

Usually the tasks set in lotto games are quite simple since the pictures on the cards and those on the board are identical. The child merely has to match images.

Simplest lotto board:

cards

Sometimes the forms to be matched get tricky and a lot of looking is required to match identical shapes. It is fun to make complex lotto matching boards and is good perceptual training for young children. An easy way to do it is to decide on a few simple forms. Using the O, X, ↑ and Δ illustrated above, I begin to combine them in increasingly complex ways to provide tricky matching tasks. For example, cards like the following can be built up:

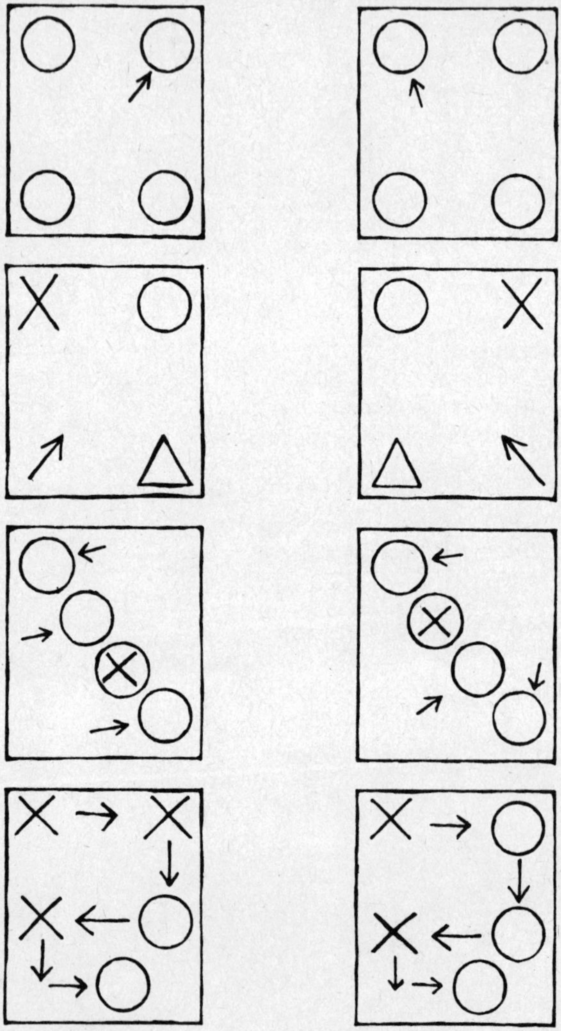

A board with forms like these but more complex can provide challenges for adults. These forms have been derived from some of the nonverbal tasks that are set in adult intelligence tests in order to discover perceptual skills: for example the ability to distinguish small modifications in complex designs. There is a series of commercial matching games for adults which consist of extremely complex forms which vary only slightly from each other. The most interesting I have seen are the Haar Hoolian Perception Games (Adult Leisure Products Corporation, Locust Valley, New York 11560) and Parker Brothers' 'Scan'. It is fun to sit down with a group of people and try to make up a complex matching game to trick others.

I have created a few matching games for phonics which use boards similar to lotto games. The boards have six squares, each with a different letter in it.

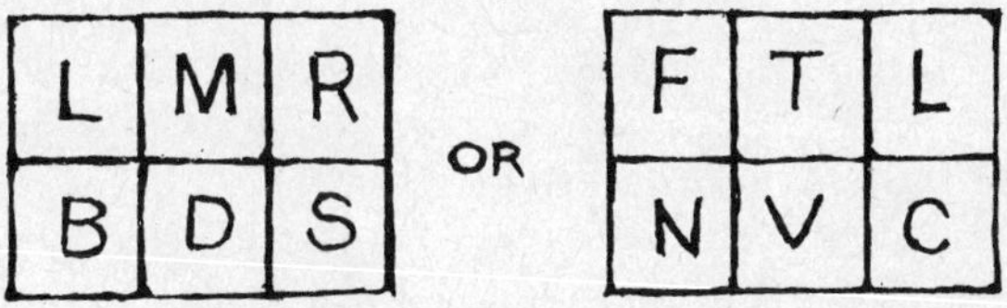

Next to the board is a little box of objects such as pins, pennies, beans, little plastic animals, cars and people, pencils, bottlecaps, crayons, etc. The object of the simplest game is to place an object on each letter on the board so that the name of the object begins with the letter it is placed on. Thus a toy car can be placed on C for car; A for automobile; V for vehicle; P for plastic.

A more complex game is to see how many objects out of the box can be fitted on the board, i.e., how many Ls, Ms, Rs, Bs, Ds, and Ss can be discovered in the box by the player with the first board.

A final variant I've played with is to ask students to fill up the boards with objects whose names *end* with the letters on the board. Thus a stamp would be placed on P rather than S in this version, which second and third graders particularly seem to enjoy.

I have played with other matching boards which don't lead

to games so much as to discussions of moral issues. For that reason I've called the boards moral matching boards. Here are some of them:

Along with these boards are dozens of cards containing pictures of animals, people, machines, products, toys, etc. Sometimes I separate these different categories; at other times the cards are all mixed together. The object of the game is to sort out the cards onto the board into the categories described. Sometimes a third category (mix) is added. There is no right way to sort the cards and it is precisely the different ways in which the world is divided into moral categories that leads to discussion and analysis of people's preferences.

Usually it is best to start with sorting animals into the different moral categories and then studying with the students the different ways in which individuals decide what is good or bad, pretty or ugly. I have found moral sorting equally fascinating to adults and to five-year-olds.

* * * *

There is a last kind of board I want to mention in this section, the vectored board on which the notation used to describe the game is built into the nature of the board. The vectored board is essentially just a piece of graph paper with coordinates drawn in and can easily be used to introduce students to Cartesian coordinates, coordinate geometry and to the reading and making of graphs.

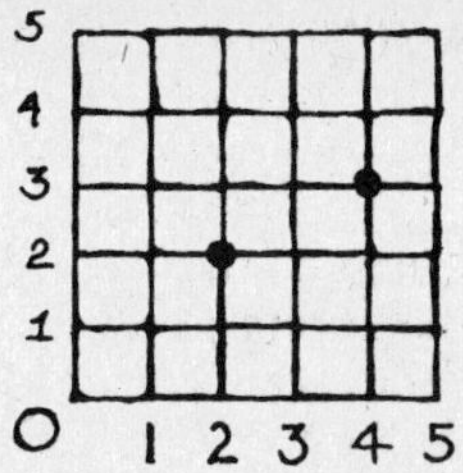

The points on the board used for play are on the intersections of the lines so that on the illustrated board the points marked out are (2, 2) and (4, 3) (observing the convention that in [x, y] the x gives the position along the horizontal axis and the y the position along the vertical axis).

The most popular game played on a vectored board is battleship. There are commercial versions of this game but it is just as easy to draw your own board. Here is a very simple version of do-it-yourself battleship:

Each player has two pieces of ten-by-ten graph paper* with coordinates drawn as follows:

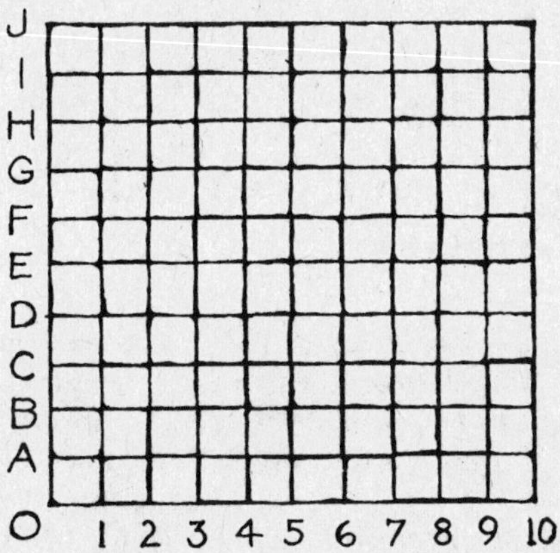

The numbers and letters are used together to avoid confusion with young children since, for example, (B, 7) and

*Or five-by-five or twenty-by-twenty, depending on the complexity of the game you want to play.

(7, B) are the same point but (2, 7) and (7, 2) are different points altogether. Sometimes simple modifications of notation such as this make a game much easier to play.

Each player has two ten-by-ten graphs. On one the word MINE is written, on the other the word ENEMY. Each player is given a navy. In the commercial game each player gets a battleship, a cruiser, a destroyer, a submarine and a PT boat. These ships are placed on the board marked MINE. The board marked ENEMY is left blank at the beginning of the game. Players are not supposed to see each other's graph sheets.

The battleship takes up five points on the grid, the cruiser and destroyer three each, and the sub and PT boat two each. They are placed on the board either horizontally or vertically so that on the following board the battleship occupies (1, H), (2, H), (3, H), (4, H) and (5, H).

What are the coordinates of the other ships?

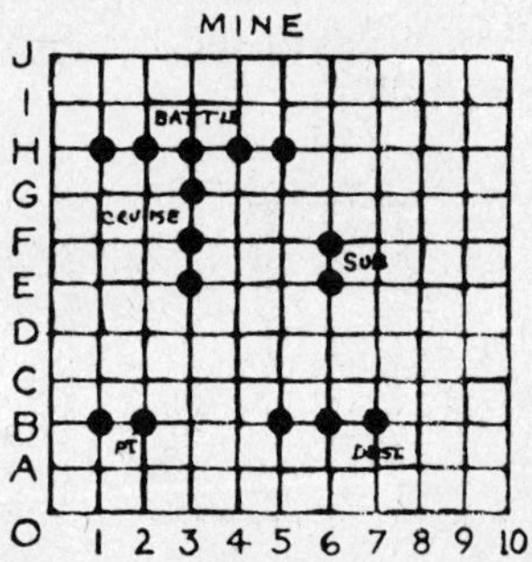

The object of the game is for the players to sink each other's ships. They take turns firing imaginary guns and calling out coordinates where the shell is supposed to land. If a ship is hit, the player hit has to call out 'hit', and records it on his ENEMY board. He also records misses so as not to repeat moves later on in the game. When a ship is sunk the player has to inform his opponent.

After five moves one player's MINE and ENEMY boards might look like this:

x represents a miss, o a hit

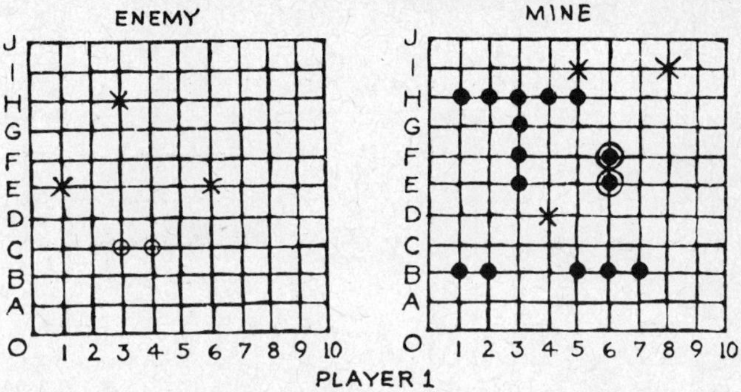

Interpreted, the graphs mean that player 1 had hit his enemy twice and missed three times but has not sunk a ship. (This tells the player where to aim next time.) On the other hand one of his ships has been sunk though the enemy has missed him three times.

Most second and third graders can master the game quite quickly and some first graders have no trouble with it. Although I know several teachers who prohibit this game because of the violence implicit in the theme it embodies, I myself don't think young people should or can be protected from all violence and I believe they should learn about self-defence. But if one objects, one could also change the theme of the game. For example, the game can be a hunt for sunken treasure. Each player puts sunken ships on his or her grid, and the object can be changed to finding each other's sunken ships. There are other themes that can be attached to this game:

— find the lost city in the jungle
— uncover buried treasure
— find the lost space ship
— find the animals in the forest, etc.

Here is an extremely simple version of a vectored game which is just like battleship though it could be called 'Find My Lost Dog'.

The game is played on a five-by-five grid. Each player knows where his opponent's lost dog is, but not where his own dog is. The dog occupies three points on the grid. The play proceeds just as in battleship and the first person to find his dog wins. Here is a sample of one player's grids after four moves:

My lost dog (enemy)

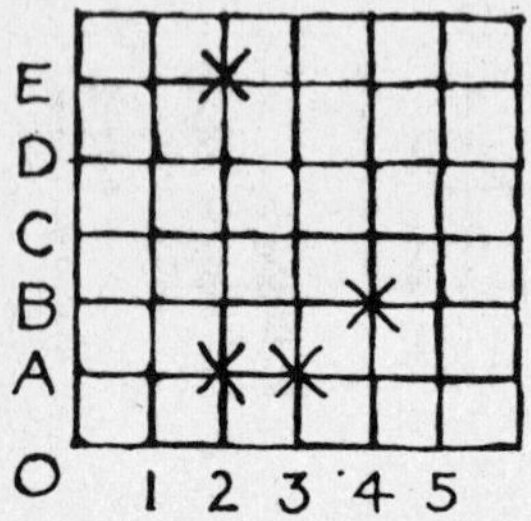

opponent's lost dog (mine)

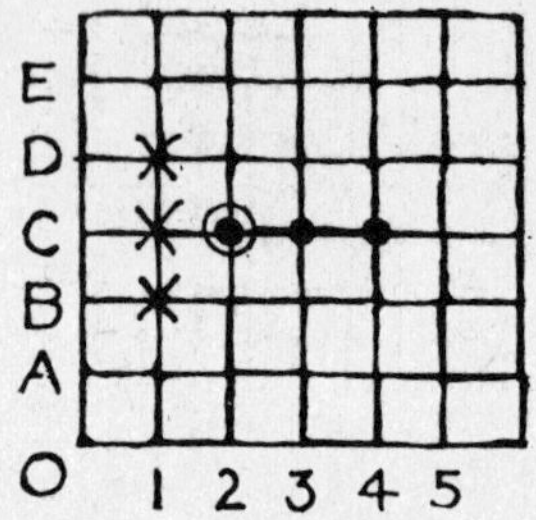

Player 1 had made no correct guesses about where his or her dog is. However the opponent guessed correctly once (2, C) and missed on three guesses.

* * * *

There is another way to use vectored boards. It is possible to play games using teams of two players each in order to get them involved in situations in which collective rather than individualistic thinking wins. I invented a simple race game for four people that illustrates some of these ideas though I think there are many other possibilities that could be tried.

The board is a twenty-by-twenty square or a thirty-by-thirty square grid. For the sake of simplicity I'll describe the game on a ten-by-ten grid. Each team of two players has a single piece. The pieces all start at 0 and the object of the game is to get to the point (10, 10) first. The board is filled with barriers so that there is no straight line from 0 to (10, 10). Players can take turns placing the barriers or a nonplaying observer can place them.

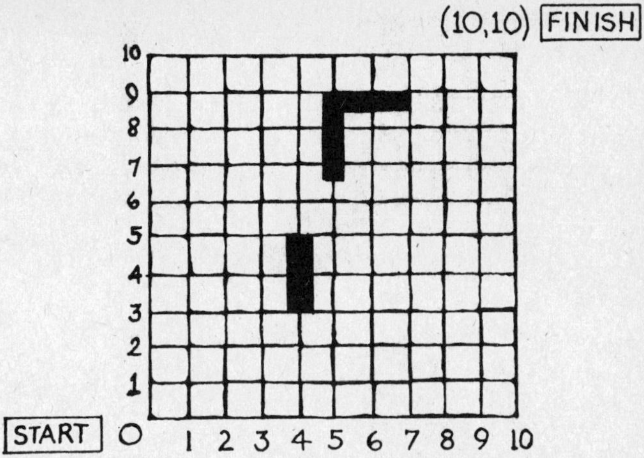

In the version of the board illustrated here there are only two barriers, one extending through the points (4, 3), (4, 4) and (4, 5) and the other through (5, 7), (5, 8), (5, 9), (6,9) and (7, 9).

On each move one player decides upon the horizontal move, the other on the vertical. No player can move any more than three squares at a time and the two players on the same team cannot communicate with each other during the game and must write down their moves before either team member discloses his move. They have to figure out from the game situation what is the best strategy to pursue assuming that the other person playing with them will also be thinking along the same strategic lines. Each move therefore will be determined by two independent decisions which will have to be made with some consideration for each other. For example, looking at a five-by-five section of the board, consider the following first moves of the two teams:

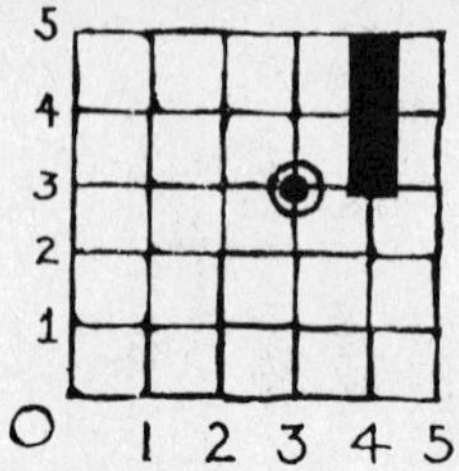

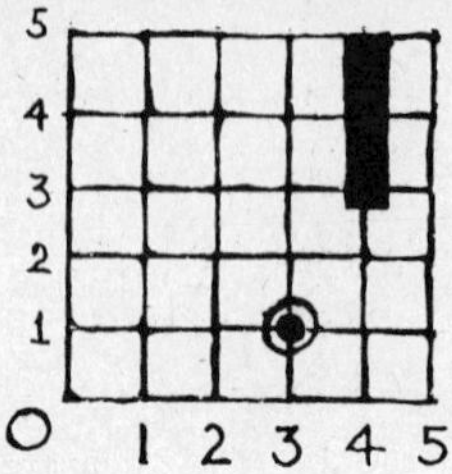

team 1:	team 2:
player horizontal 3	player horizontal 3
player vertical 3	player vertical 1

In the first drawing both players on the team went for distance and ran smack into a barrier. By choosing to move to (3, 3) they got trapped. In the second case the player moving in the vertical dimension, anticipating a long horizontal move, kept the vertical move down in order to get around the barrier. It is precisely this element of deciding both how to get to the goal and how your ally will figure out the same problem that makes this game valuable for developing a sense of cooperative functioning.

Of course no game is magic. People might learn to act cooperatively when they are playing and still cut each other dead when the game is over. Often the ability to cooperate depends upon the stakes of the game. However, even though games do not by themselves play a major role in changing people's behaviour towards each other, they do help a bit. The more people can become accustomed to playing with each other the less strange the idea of working together might seem in the rest of their lives.

3 Pieces, moves, promotions, and captures

There are two main varieties of game pieces: those that have some meaning beyond themselves (such as the kings, queens, knights, etc. in chess) and those that merely serve as marks or tokens (such as the Os and Xs in noughts-and-crosses). There are also some pieces that may once have had meaning (such as the pieces in chequers which originally were all called queens) and lost it over the course of time as the game spread to cultures where the social analogies embodied in the game were no longer significant.

There are some very simple games that are played with tokens and use no board at all. The simplest might be called even-odd. It consists of laying out a number of tokens on any flat surface. Pennies, poker chips, wooden matches, stones or any other kind of markers can be used. The game, which usually is played by two people but can be extended to three or four players, consists of the players taking turns picking up tokens. On each turn one or two tokens can be chosen. The goal is to take the last token (or in another variant force your opponent to take the last token). The game is played with either an even or odd number of tokens, and strategy for the game depends upon which is used.

Suppose the game is played with three tokens and winning consists of taking the last token (a sitution too simple to challenge most people but easily analysed). The first player can take one or two tokens. If he or she takes one token, the second player can take two and win. If the first player takes two tokens, the second player takes one and wins. This game will always be won by the second player if mistakes are avoided.

A game with four tokens changes the situation somewhat. If the first player to move takes two tokens there are only two left and the second player can take those and win. However, if the first player takes one, then three are left and there is essentially a three-token game left to play which reverses the order — player 2 taking first and therefore losing as was outlined above.

This is an important argument to follow. Essentially what it says is that by taking one token on the first turn the first player leaves the second player in the position of being on the losing end of a three-token game. A diagram illustrates that situation:

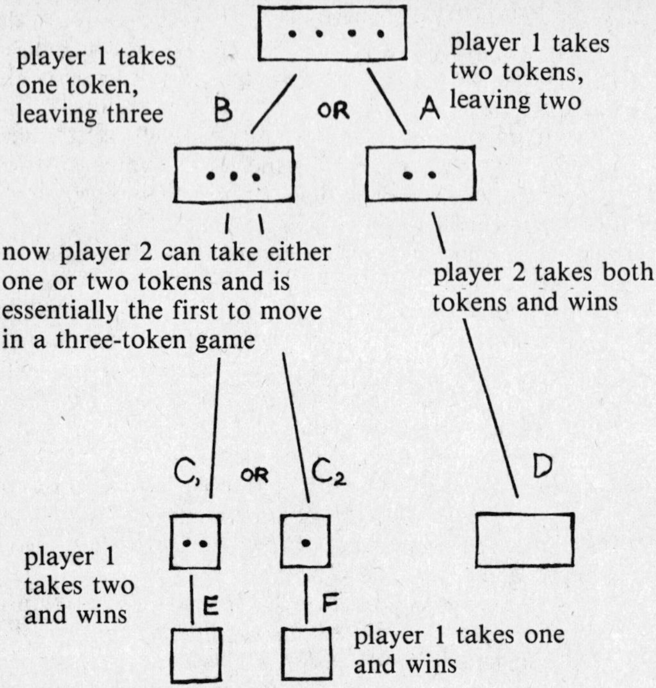

Thus in a four-token game if the first player takes one piece he or she should always win.

Now try to generalize these simple results. What would be the best strategy for playing a six, eight, nine, ten, fifteen or seventeen token version of the game? Are there any rules of strategy that can be discovered so that you maximize your chances for winning? I've seen seven- and eight-year-olds solve this somewhat sophisticated problem with games involving between five and one hundred tokens and discovering through the game some of the differences between odd and even numbers.

A slightly more complex version of this token game is called NIM. The game, which was made popular in the film *Last Year at Marienbad*, is sometimes referred to as the Marienbad Game.

The game consists of laying down tokens in an order

similar to the way bowling pins are placed. Any number of rows, each one token longer than the last can be used. A five-row game would begin with the following configuration:

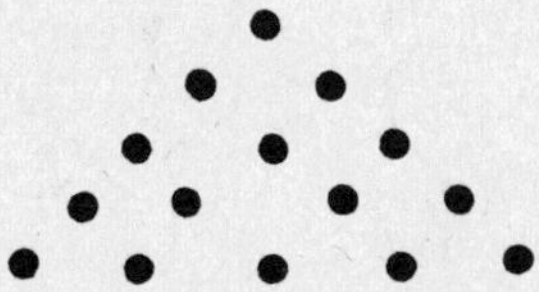

The rules of the game are that the players take turns picking up any number of tokens from a single line during their turn and the person taking the last token wins (in some versions the person forced to take the last token loses). Thus the following are legitimate first moves:

1) two from row 2, leaving:

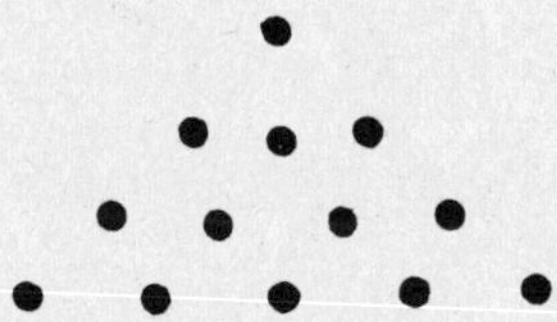

2) three from row 5, leaving:

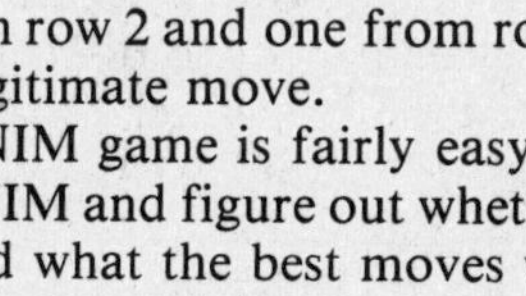

Taking one from row 2 and one from row 5 in a single turn would not be a legitimate move.

A three-token NIM game is fairly easy to analyse. Try to play three-token NIM and figure out whether it is better to go first or second and what the best moves would be.

Then extend your strategy to three-row, six-token NIM:

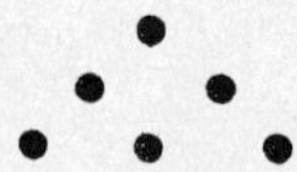

and after that, if you aren's bored by the game or the ananlysis, try to develop an overall strategy for this simple NIM game and test it out.

There is a commercially-available game called 'trinim' which is a sophisticated version of NIM and worth experimenting with. It is available from Wiff 'N Proof, PO Box 71, New Haven, Conn. 06501.

* * * *

As interesting as NIM is, I prefer games that go beyond pure strategy and connect with living forms. If chess pieces were merely numbered, the edge of the game would be lost for me. Capturing the king, mobilizing the knights, developing a strategy for holding territory or trapping a queen make the game magical and not merely an interesting thinking exercise.

Even the simplest chance games can be dressed up to interest young children. By providing boats or characters from stories like Pinocchio or Winnie the Pooh simple race games are more fun. There is something to connect with in the pieces, some life to the game. For young children and many adults too, games have to embody stories or adventures to be interesting. I have been trying to figure out a way to develop interesting strategy games for and with young children, games which do not merely depend upon spinners or dice but which are not as complex as chess. As is often the case, my children and their friends have helped me to discover some solutions. One day I noticed that my daughter Tonia and two of her friends were playing with small plastic animals and a chess board. They had set the animals up opposite each other and were moving and jumping them around the board. On one side there were two elephants, a cheetah, a mouse, a snake, a polar bear and two poodles. The other side had three rhinos, two horses, an owl, a parrot and a cow. I watched the kids playing. When they began the game there were no apparent rules. They moved the animals around the board, gave them voices, and occasionally knocked them against each other. Then patterns began to develop. The owl went after the snake, the snake went after the mouse, which tried to scare the elephant. The children began talking about how the animals should move on the board — whether the snake, bear and elephant should all have the same mobility. They talked about which piece could capture the other. Cows shouldn't be able to capture elephants — or should they?

After a while the children became bored and put away the board and animals. But a few nights later I returned to the issue with my daughters. We sat with a checker board and a box full of little plastic animals. Each of us chose two animals, put them on the board and played around with how they might move. We talked about the animals themselves, about where they lived, how they actually moved, who their enemies were, what their prey was. Then we translated this analysis onto the board and moved the pieces about, inventing games as we went along.

Since then I've fooled around with games using animals, people and vehicles with several adults and young people. We've talked about or studied the lives or mechanical functions of the pieces we were using and then tried to translate some of this information into game terms. The initial exercises were to make up moves and seizures, then to develop combinations of pieces and boards and make games. Next year when I teach in a K-3 grade class we'll actually construct samples and play the games we create as well as teach them to other classes.

Here are some sample animals and moves:

1. The CRAB scuttles sideways and captures with its claws. Therefore logical moves for it in a game would be sideways. However, if it only moved sideways it would be stuck in one part of the board. I discovered this immediately by putting the crab on a chequer board and experimenting with the proposed moves. The crab in the diagram could move any number of spaces sideways but is trapped in the eighth row.

A simple modification gives the crab greater mobility — one space forward or back and then any number of spaces to either side. The potential moves can be represented as follows for a piece placed in the centre of a board:

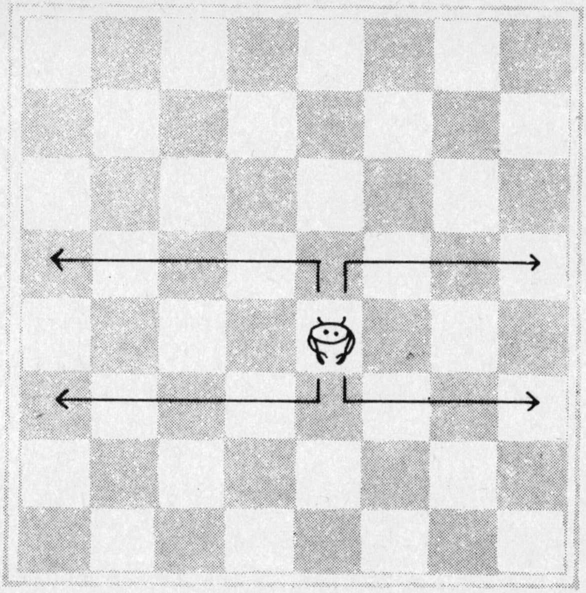

Crabs capture with their claws. Visually one can imagine a crab scuttling along sideways and then making a quick forward thrust to capture:

Therefore a further modification of the crab's movement can provide for capturing opposing pieces. The crab can move (1) one forward or back and any number of spaces to the side, or (2) move any number of spaces to the side and then one forward or back to capture. Capture can only be made on this last thrust.

The following capture of a fish illustrates this second type of move:

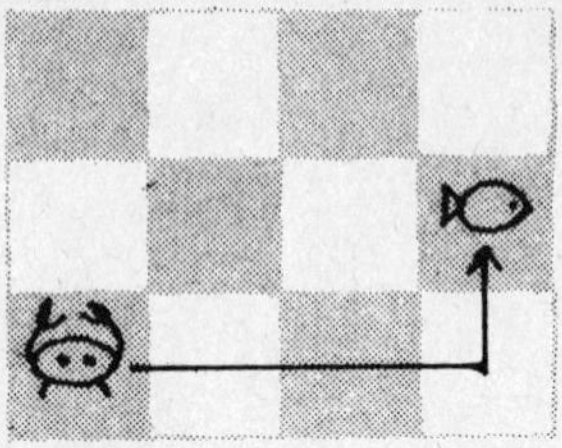

The crab's move is not unique in the history of games — it can be looked upon as an extension of the galloping move traditionally given to the knight in chess.

2. BIRD moves are not so easy to create. There are all kinds of birds — parrots, canaries, swallows, owls, pelicans, pheasants, crows, hawks, jays, etc. Their natures differ substantially — their flight range, prey and manner of attack lead to a wide variety of movement and capture patterns. The ability to fly, common to most birds, can be interpreted in game terms as an ability to jump over pieces without taking them. In chess the only piece with this power is the knight, and his moves are extremely restricted.

Also, using flight to define the birds' moves lends itself to differentiating the moves of different types of birds. For example:

a) the crow, which flies straight, can move up to four squares vertically or horizontally and can jump over intervening pieces.

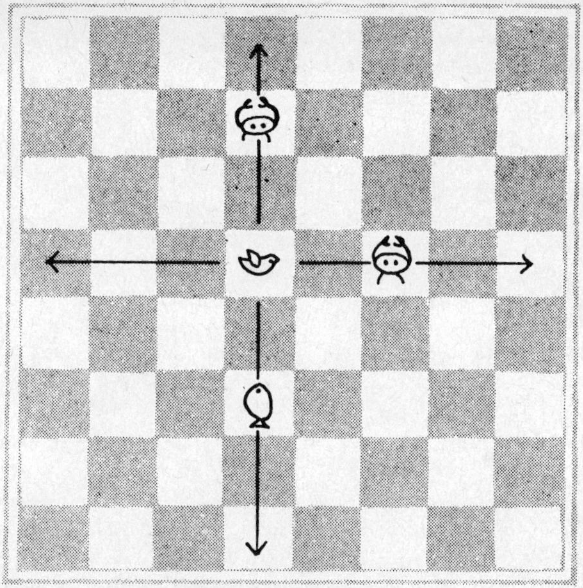

straight as the crow flies

b) the jay can move three squares horizontally, vertically, or diagonally and also has the ability to jump:

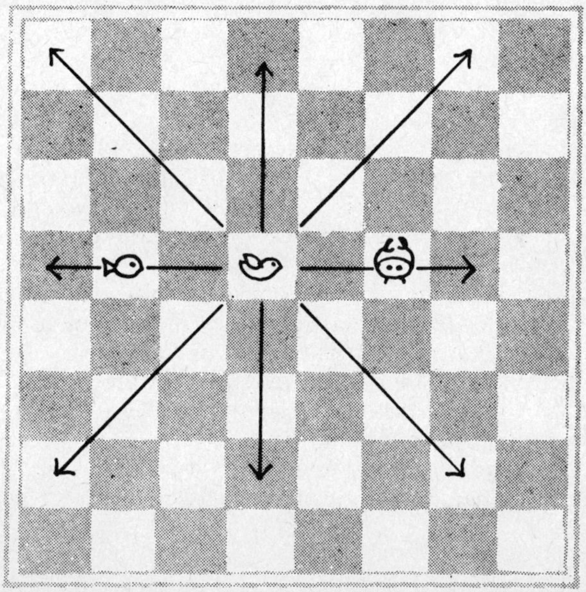

c) the hummingbird makes darting movements. therefore its moves can be either one or two spaces horizontally, vertically or diagonally.

Two different hummingbird moves:

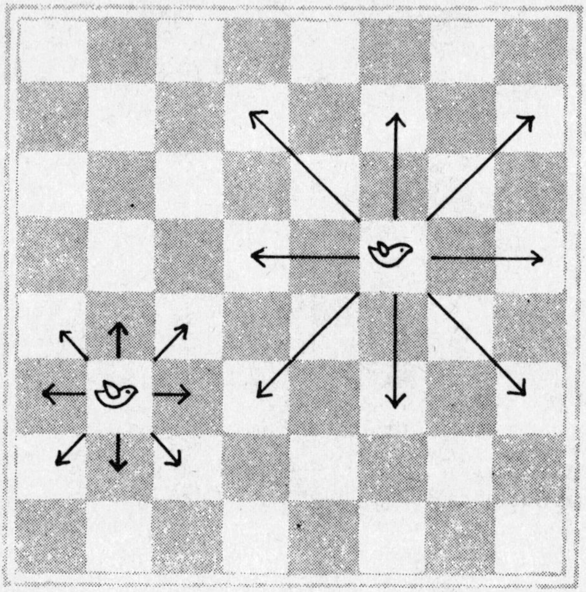

Given the range and variety of bird moves that can be developed from these simple concepts it probably makes sense to make a uniform procedure for birds capturing other pieces. A simple rule would be that the bird captures what it lands on. This gives birds a lot of power and could of course be modified.

3. ALLIGATORS strike people as sneaky creatures. They move very slowly, they dart out, they seem to be asleep but are constantly aware of the postion of their prey. They also change directions a lot — flit over in the water, turn quickly when their prey tries to escape. Some of these characteristics can be incorporated into moves. Suppose the alligator moves with a special rhythm — slow, fast, slow — and that it also takes the alligator one move to change direction. This can be

specified in the following way:*

Rules of alligator movement:

a) alligator moves one square forward to begin with;

b) on its second move it can go any number of squares in the same direction (but not jump over intervening pieces);

c) on the third move it can move only one square in the direction it faces;

d) throughout the game the sequence must be maintained — one square; then any number of squares possible in the direction the animal faces; then one square again; etc. The sequence can be represented as 1-x-1-x-1-x, etc.

There is one additional rule:

e) the alligator at any move can change direction by turning on the same square to face a new direction (vertical or horizontal) in which he wants to continue. This counts for a turn, and on the next move the sequence is resumed where it was left off.

Here is a sequence of alligator moves:

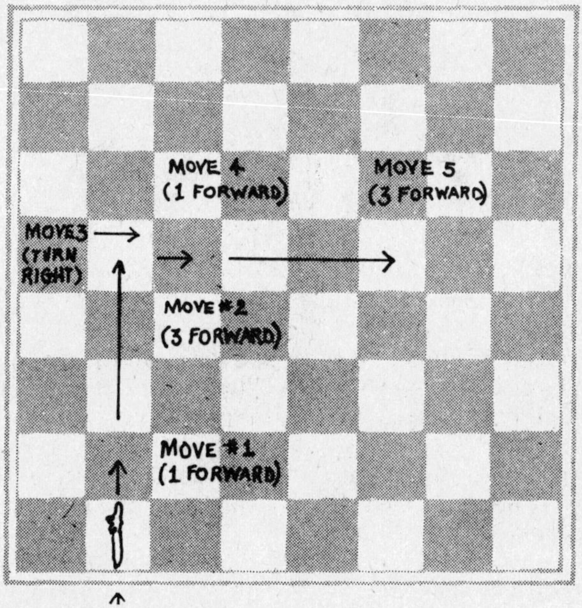

*Of course there are many other ways these characteristics can be turned into moves.

A simple way to deal with capture is to allow the alligator to eat whatever it lands on.

Keeping track of the sequence of alligators might be a problem. A small counting device like one used in a counting knitting stitches can be used.

Here is the initial position of a game that might be played on an ordinary chequer board with the animal moves just described:

Each side has two blue jays, two crabs and two alligators. The object of the game is to capture all of the opponent's pieces. I have played this game a few times with young children. One thing that seemed clear was that the board was too large. Therefore I cut the board down and the resulting six-by-six board seemed to make the game quicker and more interesting.

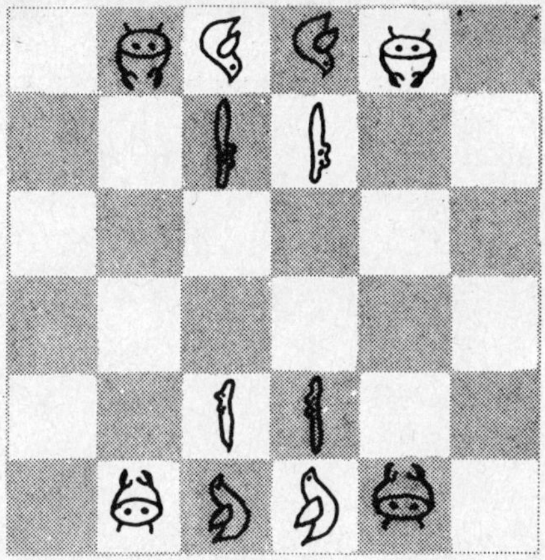

Children seem to love games like this because they love to move the animals involved and to pit them against each other. In my classroom (a K-1 class) the children play a lot with a box of animals. They use the animals to act out their own fantasies about being wild, about living on their own, dealing with enemies and making friends. Free-flowing fantasy using animals seems a natural way for four-, five-, six- and seven-year-olds to explore their emotions. However, these children also seem to have an impulse to organize these fantasies. They want to play with one another through the animals, and animal strategy games provide one way for children to play at conflict without actually fighting.

There is a simple animal game that can be played on a three-by-three board. Is there a strategy you can figure out for the best possible way to play this game using the rules for moving and taking described above?

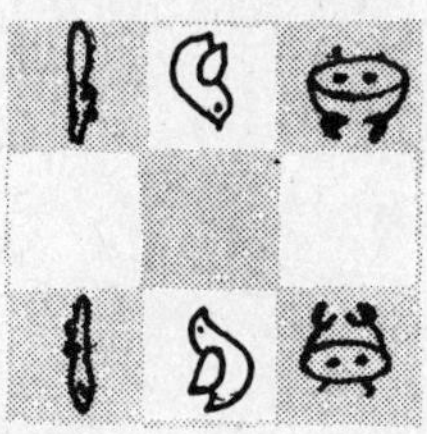

Of course there are other creatures that can be used to create games. As practice try to create moves for the following objects, using a chequer board as the field to move on (also ask your students to try):

lobsters	politicians	fish:	lovers
monkeys	cars:	sharks	students
snakes	Fords	guppies	teachers
tigers	VWs	trout	whales
cops	Porsches	a pack of wolves	dinosaurs
rhinos	fire engines	chickens	trucks
horses	police cars	rowing boats	sailing boats

Also try to create moves of two pieces at a time in games that involve fundamental oppositions: prisoners/guards; students/teachers; madmen/psychiatrists; friends/enemies; children/adults.

The parent-child opposition suggests a whole series of moves. One example is of a game played on a five-by-five board with one side consisting of two adults and the other of five children. The adults can move two squares in any direction, horizontally, vertically or diagonally. The children can move one square in any direction. Both sides capture merely by ending their move on a square occupied by an opposing piece. Here is a diagram of a possible starting position of the game:

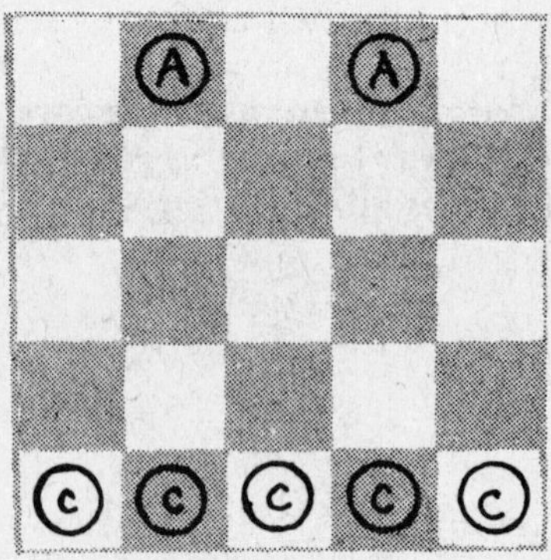

* * * *

For the sake of convenience I've been using chequer boards for the animal and people games described above. This isn't necessary, and a group of third-graders suggested that a board be made for an alligator/bird/crab game that represented the kind of terrain each of the animals live in. Thus some trees, a river, and an ocean were drawn on a large piece of graph paper (with one-inch squares which provide a convenient grid for the creation of game boards). The board the class arrived at after a few attempts looked like this:

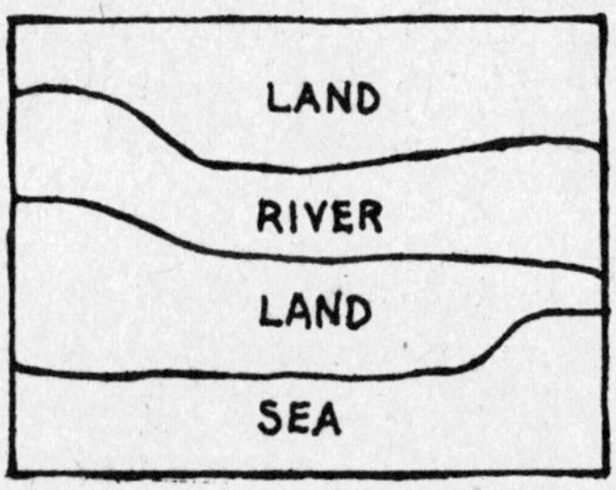

The students tried to create equal areas on the board of river, sea and land and eventually did so by counting the number of squares on the whole board, dividing by three and then redrawing the division so that one-third of the squares represented each area.

Another problem arose. The crab lived in the sea, the alligator in the rivers and birds on land. Could the animals move out of their elements? The obvious answer had to be yes or there wouldn't be any interactions in the game. So the students decided to make safe spots, nests in each area for creatures whose element it was. Thus there was a spot in the sea where crabs were safe and could not be captured, nests on land for the birds and marshes for the alligators. The board when finished looked like this:

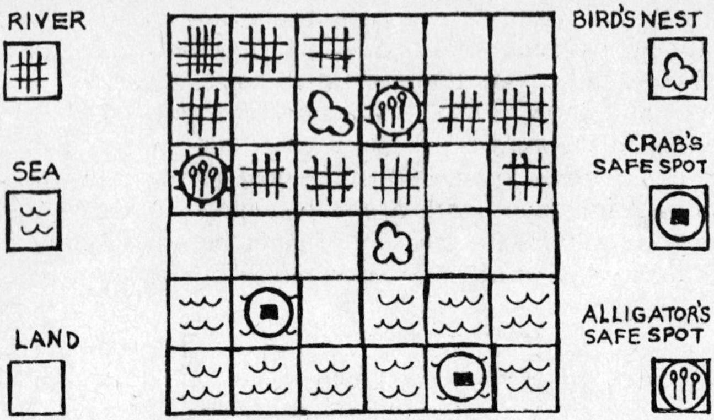

Each side had one crab, one alligator and one bird, with moves and captures according to the rules listed above.

The game starts with the creatures placed in their safe spots. The goal is to chase and capture your opponent's animals.

Other boards can be created for games using the same pieces. Grids with circles or pentagons or triangles can be experimented with in order to discover which configuration of forms gives the kind of mobility desired in the game. It is probably a good idea for people interested in creating games to have the widest variety of graph paper obtainable.

I find graph paper a bit limiting and asked a friend, Jean Doak, to make a series of circles, triangles, squares, pentagons, hexagons, heptagons, and octagons for me out of wood. Each block is approximately four-and-a-half inches in diameter and three-quarters of an inch thick. Each shape is a different colour — the triangles red, the squares green, etc. I experiment with the blocks, putting them together in various configurations to see what different game boards might look like. There are twenty-five blocks of each shape, enough to give me an idea of what a playable board might look like.

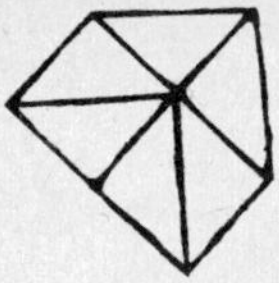

triangular grid

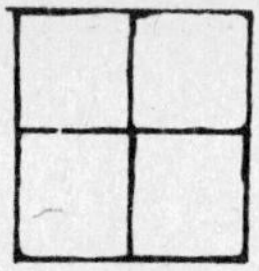

square grid

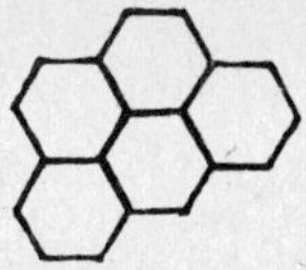

hexagonal grid

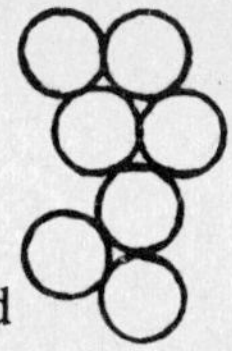

circular grid

mixed grid using squares, triangles, and circles

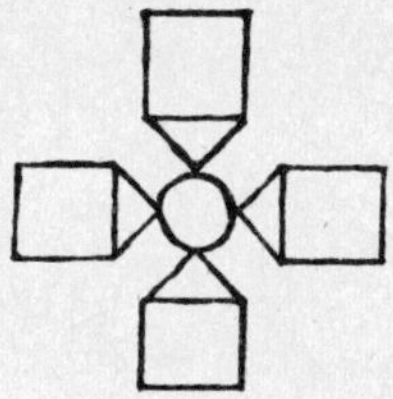

Recently while looking for ways of extending the uses of my blocks I came upon a game in which the pieces become their own board. It is called Inter-Dependence: A Game of Ecological Relationships and is produced by Dolphin Educational Resources, PO Box 22068, Seattle, Washington. The game consists of arrows and circles. On the arrows are written relationships which exist between different forms in nature:

There are also a few blank arrows, wild cards as it were.

The circular cards have picture on them along with words describing what the pictures represent. Different natural forms appear on them:

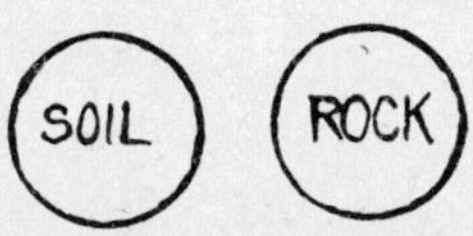

people
mountain lion
plankton
rain
berries
algae

tree
fox
bird
city
sun
etc.

In the course of the game, circles ar joined by arrows according to relationships that actually exist in nature, such as:

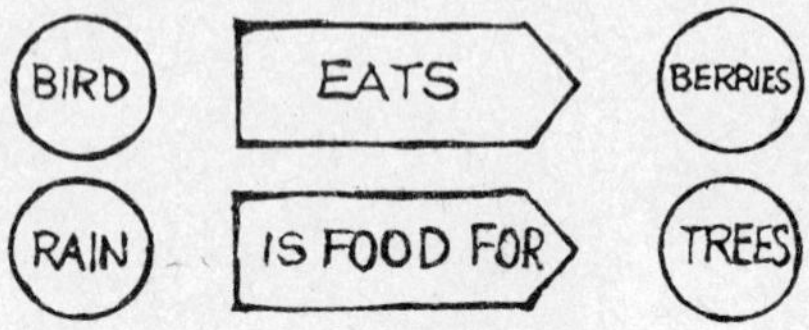

There are also blank circles. Here are excerpts from the instructions included with the game as well as an illustration the game makers included to show how complex some of the cycles can become:

> *These rules are intended to be only suggestions. You will probably have fun inventing other games to play with your cards.*

To play the game:

- After shuffling the cards, remove any one round card and place it face up in the middle of the table.
- Divide the remaining round cards among the players.
- Place the long cards face down, in a pile within reach of all.
- Each player in turn draws a long card, and using any card from his or her hand, completes a relationship, building out from the centre card.

Remember:

- Any number of relationships may be connected to a round card at one time.
- More than one relationship may exist between two round cards.
- Any player may question a relationship played by another. If the relationship in question cannot be settled by explanation, the majority should decide if it is acceptable.
- There are blank cards, both kinds; to be used for anything a player wants.

Here is how a typical game might grow:

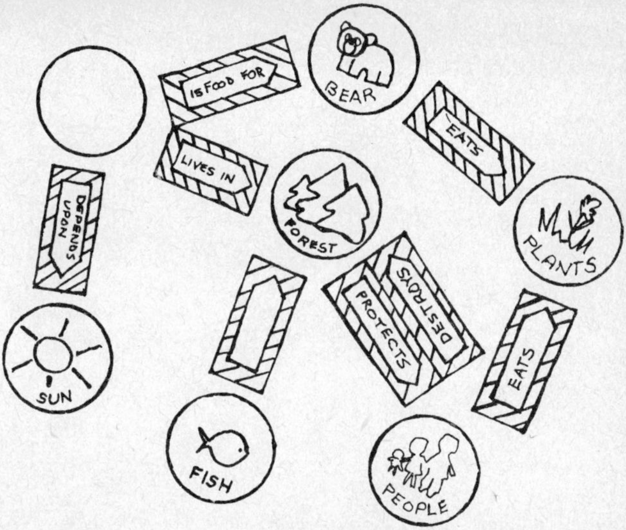

Playing around with Inter-Dependence gave me some ideas on how to extend the concepts involved in such a game where the board and pieces were identical. I asked Jean to make some arrows for me out of wood and got several different forms of arrows:

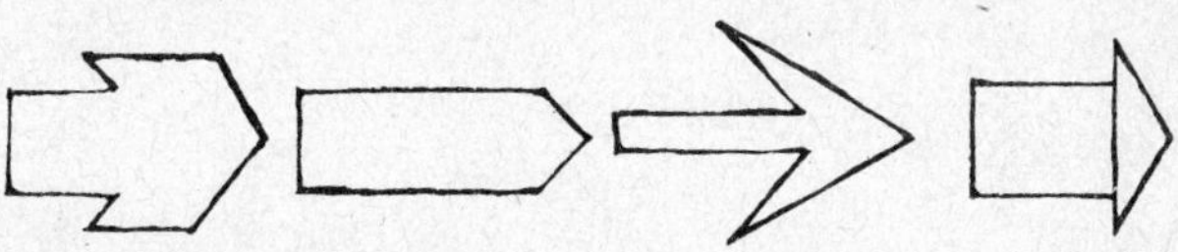

Using the arrows and wooden shapes, I began experimenting with relationship games. The different shapes lend themselves to different forms of routing things. For example:

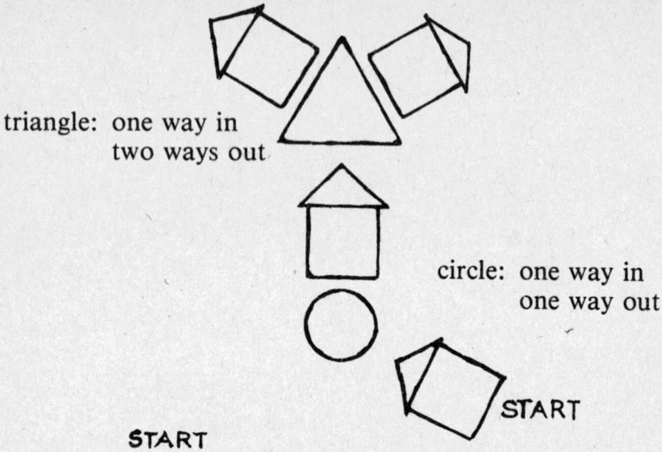
triangle: one way in
two ways out
circle: one way in
one way out
START

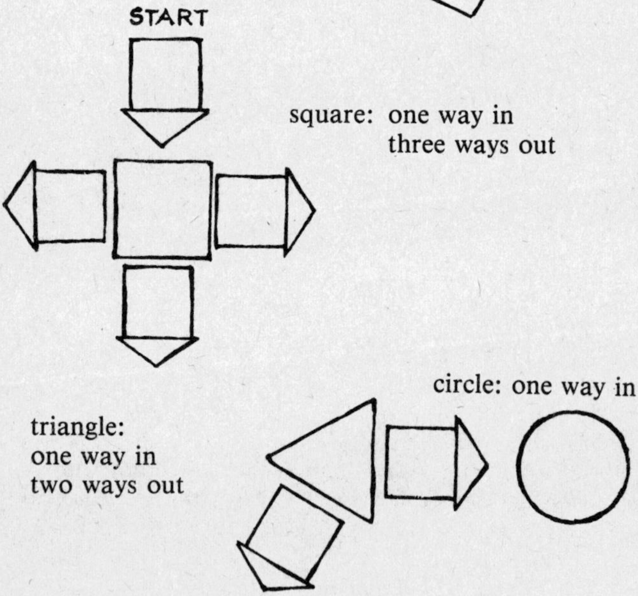
START
square: one way in
three ways out
circle: one way in
triangle:
one way in
two ways out

My daughter Tonia made up a board using one square and a number of circles:

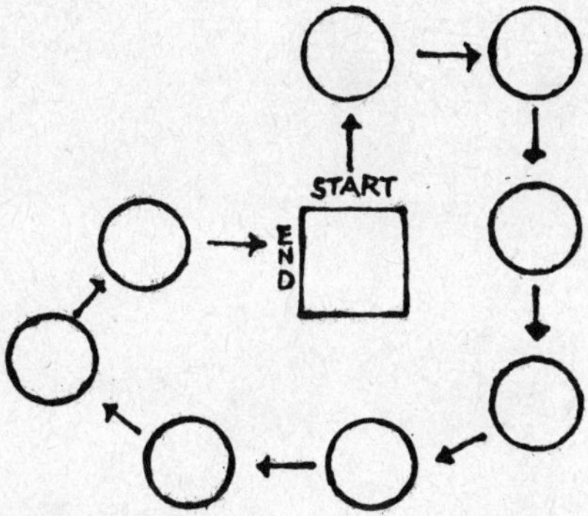

Then, using two plastic horses placed at the start and a spinner to determine how many blocks can be moved at a time, she raced the horses from start to finish. She played the game by herself because, as she said, winning games made her too nervous. It was easier for her to have the plastic horses compete against each other than to have two players competing using the horses as their surrogates.

There are ways that the blocks can be worked in with spinners to develop games with more complex routes than the simple race game structure that Tonia made. For example, using a simple 1/2 spinner

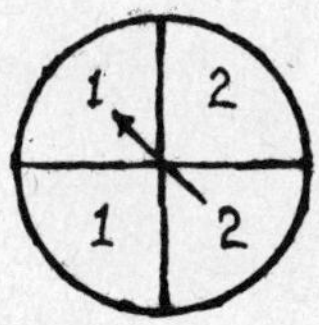

and labelling the faces of the blocks 1 or 2, the following complex race track can be used with Tonia's horse race game:

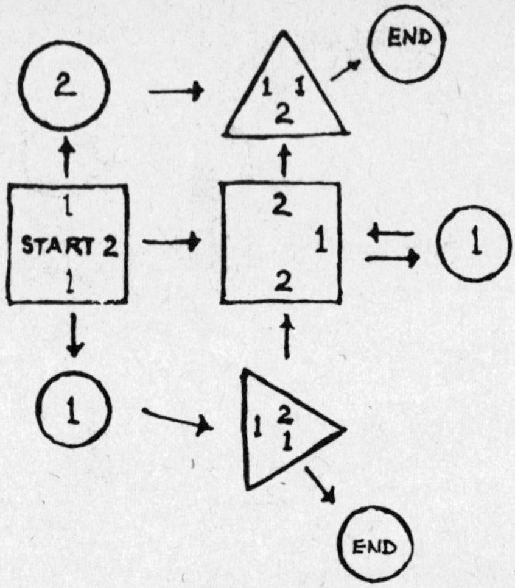

On this board there are two end points. There are also a number of different routes that can be taken to get to these points.

The pieces begin on the side marked 'start'. If a 1 comes up on the spinner the player can move either to the left or right (the sides marked 1) to the next block (the circles). If a 2 comes up the piece moves straight ahead to the other square block. A piece can only move if the number coming up on the spinner is the same as one of the numbers on the block it is on. Thus if a piece is on circle 1 it can only move when a 1 comes up. In this game there are times when the player can choose alternate routes. The board has to be analysed so that the shortest route to the end is taken. This introduces an element of thinking into the simple chance-determined race game. It is easy to imagine much more complex routing using six- and seven-sided blocks.

I want to talk about only one more aspect of game moves in this section — promotions and demotions. In chess a pawn reaching the opposite end of the board can assume the power of any piece on the board. In checkers a piece reaching the last rank is promoted or kinged so that it can move backward as well as forward.

There is a more interesting case of change of status in Sho-gi, a Japanese chess-like game which is the most sophisticated and difficult strategy board game I've encountered. In Sho-gi all the pieces are of the same colour and are shaped like arrows. When the board is set up it is possible to tell which of the two players has a particular piece by the way the arrow points. I have devised a simplified, chess-like version of the game.

I use a five-by-seven board with each side having ten pieces — five pawns, two bishops, two rooks and a king — set up in the following way (each piece is a small arrow with a letter written on it indicating its status):

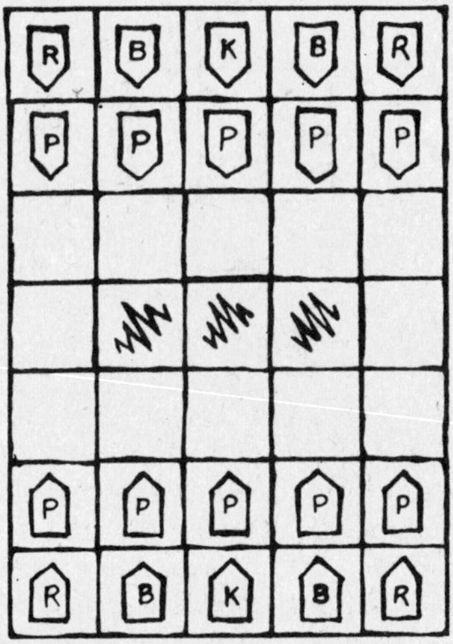

There are three special squares marked in the middle of the board. These are promotion squares. If bishops land on these squares they are promoted to rooks. To facilitate this happening I've labelled the under side of the bishop pieces with an 'R' (as in Sho-gi, where a number of pieces are labelled on both sides and are flipped over when they land on promotion squares). The pieces face the direction of the opponent, away from the player who commands them. In the game I've invented, other than the rule of promotion and one other rule, everything else is exactly as in chess.

The final rule, and the one that makes Sho-gi-type games unique, is that when you capture a piece it becomes your own piece and can be re-entered into the game on any turn you please. The captured piece is simply placed on an empty square facing the opponent. This introduces a whole new element of strategy into the game since the outcome is determined not merely by the present disposition of the board as in chess but by how captured pieces might be re-entered in play, changing the whole situation.

Thus in the following situation the pawn can capture the bishop and

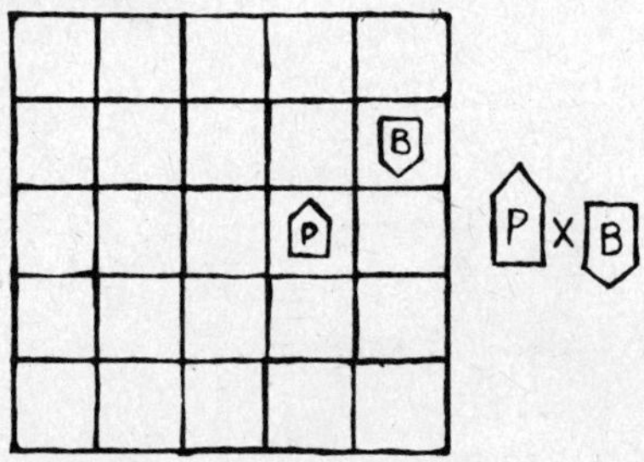

on some subsequent move the bishop can be re-entered in play on the pawn's side.

Ideas from this game can be introduced to all kinds of other games. Squares can be designated promotion or demotion squares, the powers of pieces can be changed in different ways in the course of a game, pieces can be captured and re-entered into play on the other side.

In this way, for example, checkers can be changed so that:

1. a piece captured can be returned to the game on the other side;

2. squares can be designated on the board as demotion squares and if kings land on them they lose their king status.

Another modification of chequers might consist of having promotion instead of demotion squares in the middle of the board.

In animal games it might be possible for blue jays to turn into eagles or be demoted to the worms they eat. There is no end to the number of variations it is possible to use in creating more and more interesting games.

4 Decision-making devices

The outcomes of some games are determined purely by chance, while those of others depend only on the skill of the players. Most games, however, use some combination of chance and skill.

The skills of game playing include almost all the abilities that have been traditionally and formally taught in school. For example, ring-toss games, marbles, pitching pennies and bean-bag tossing all require 'hand-eye coordination', and scoring them involves addition and subtraction skills. Chess and other strategy games require the ability to think through several alternatives simultaneously and to develop overall plans and then devise specific steps to carry out these plans. To play games like noughts-and-crosses you must learn inductively, i.e. from playing absorb ways to avoid repeating mistakes.

Even simple chance games use scoring systems that encourage practice of arithmetic skills. Young children love to experiment with chance and luck; to toss dice or pick cards or pick names out of a hat. There is an element of excitement in the uncertainty of chance games as well as safety in them, since everyone has a chance of winning, even those children who habitually lose at games of skill and strategy. I've noticed a lot of students who keep away from chess and chequers because the competition creates too much anxiety yet play complex chance games obsessively. Losing is much easier to take if the outcome is out of one's hands.

One of the most common chance devices is the spinner, which consists of an arrow that spins around a board and stops at random. The simplest spinner boards consist of equally divided and numbered sections which can be used to determine moves or points. Not all spinners are simply numbered. Some use colours, written directions or letters:

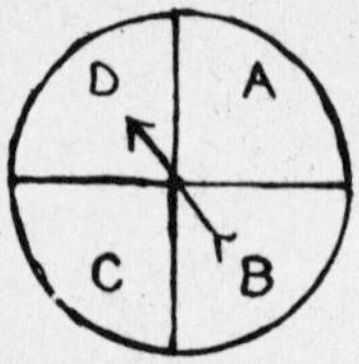

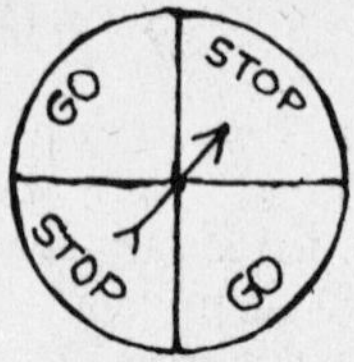

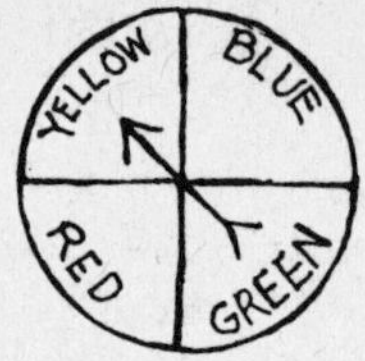

A simple game using a colour spinner could be played on a race-type board with the colours on the spinner repeated on the board squares:

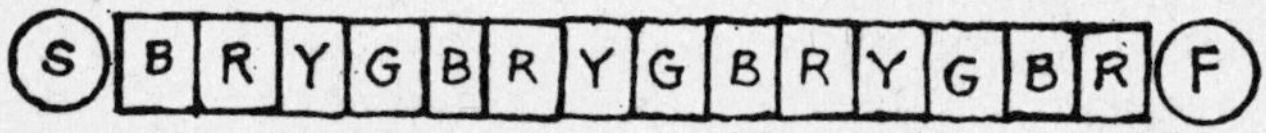

B	Blue	G	Green
R	Red	Y	Yellow

The players put their pieces on the start cicle, spin the arrow, and move to the colour indicated. The first player to reach the end circle wins.

The game can be made a bit more complex and interesting by modifying the spinner:

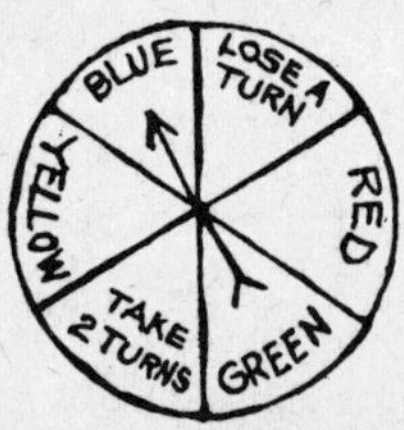

I have named one of my favourite spinners 'the oracle':

This spinner answers questions, and it is possible for games to be developed using questions randomly answered. For example, a pack of cards can be marked with questions such as:

— Do I have to kiss the person on my left?
— Do I have to spin around until I'm dizzy?
— Do I have to pretend I'm a chicken?

Each player takes a turn picking a card and consulting the oracle. This noncompetitive game is both a lot of fun and a good way for beginning readers to get a lot of practice.

Sometimes weighted spinners give games more suspense or make them more interesting or quicker to play. If one wanted everyone to act out a card, a weighted spinner with three 'yes' sections and one 'no' section would get people involved more quickly; if suspense was desired a three 'no' and one 'yes' spinner would be better to use.

It is easy to make spinners out of cardboard and brass paper fasteners.

Do-it-yourself spinner

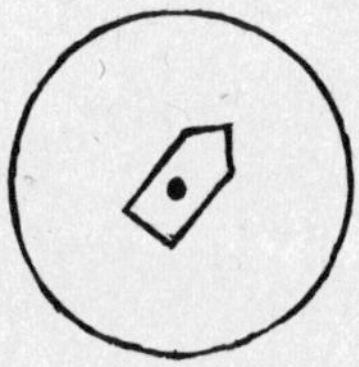

Here are some spinners that students made:

1. for an adding and subtracting game:

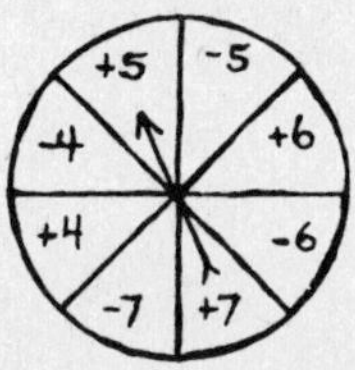

2. to determine teams and groups to play cooperative games:

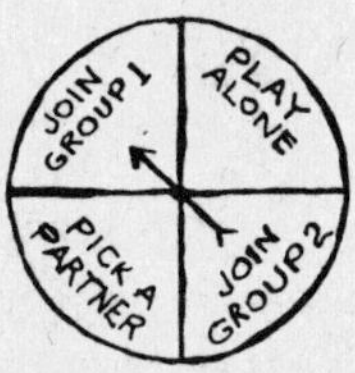

3. for a game involving moving about the playground to reach certain safe places (notice how this spinner was weighted:

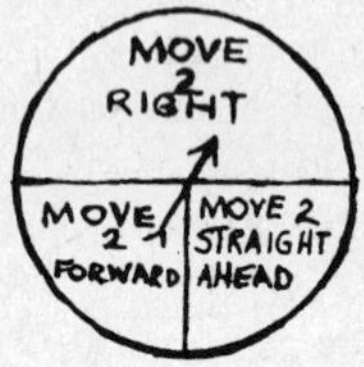

* * * *

Dice are used as frequently in making game decisions as spinners. There are also a whole range of games played with dice themselves (craps being only the most familiar), as well as a variety of different types of dice. Therefore it is useful to create a notation system to describe the different faces of a dice. The simplest way to do this is to take the physical dice and unfold its faces on paper.

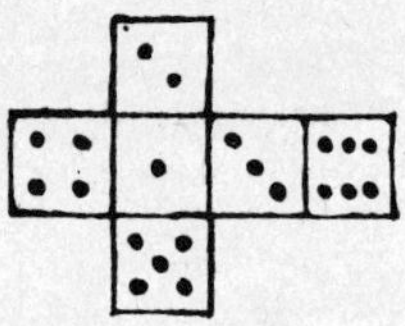

Then number the faces according to the number of dots that appear on each face:

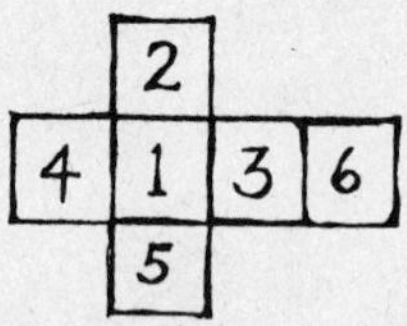

You can use other ways to relabel the faces of a die — e.g., label the sides A, B, C, D, E, F or use the colours yellow, orange, blue, green, red and black for each face.

There is a simple way to make simple dice race games more complex and to teach the concept of negative numbers at the same time. Instead of using two white dice, use a red die and a white one. Each player castes the dice and calculates the number of squares to move by subtracting the number on the red die from the number on the white die.

Thus if a player casts a 4 on the white die and a 6 on the red die the move consists of 4 — 6 = — 2, or two moves backwards. It is not even necessary to explain negative numbers to students. It suffices to explain that the white die determines moves forward and the red one moves backward. Later on when the concept of negative numbers is introduced (and it's not hard to do it in the first grade) the mixed set of dice can be referred to in order to help the students understand what you're talking about.

Add a green die and the determination of moves can get more complex. For example, the number of squares to move can be (red — white) x green.

This might seem a bit laboured but it really isn't. When I began teaching arithmetic to fifth graders, dice made me nervous. I was afraid to bring them into the classroom because parents and supervisors might think I was encouraging gambling. The kids wouldn't do simple addition drill, but they loved to play craps. I solved the problem by bringing in sets of red, green and white dice. The kids were allowed to play with mixed pairs only if they used combinations of operations. They had no trouble playing subtraction craps or subtraction and multiplication craps.

I've noticed that even in chance games some children have all the luck and others have none at all. Habitual losers naturally tend to avoid playing chance games and often feel afraid of them. In order to avoid having a group of losers in my class I often change the goals of the game. One day the highest score may win, another day the lowest. Sometimes I ask the students to play in teams so that the highest or lowest collective score wins. This evens things out a lot.

Not all dice are numbered. I like to play around with labelling blank dice. (If you can't get blank dice, blocks with gummed labels will do.) For example, it is possible to weight the dice so certain combinations are more probable than others:

	MARKINGS			
FACE	*1*	*2*	*3*	*4*
1	A	5	no	blue
3	B	2	yes	blue
3	A	2	no	red
4	B	1	yes	red
5	A	1	no	yellow
6	A	1	no	green

On die 1, A has a four in six (4/6) chance of coming up and B a two in 6 (2/6) chance. On die 2, 5 has a 1/6, 2 a 2/6 and 1 a 3/6. On die 3, 'no' is more likely to come up than 'yes' by 4/6 to 2/6 and on die 4 blue and red are equally likely to come up two in six times while yellow and green have a one in six chance.

Weighted dice can be used in certain games. For example, a die with marking number 2, where 5 has only a one in six chance of turning up, could be used in race games instead of a regular die. This makes the game more suspenseful because most of the moves will be one or two spaces. When a 5 comes up a player who is behind can suddenly leap ahead. In the same way someone who has thrown a 5 and gone ahead can easily come upon a string of 1s and fall behind. This makes for more major changes during the game than using a regular die where each number has an equal chance of occurring.

Weighted dice are useful in introducing students to the concept of probability. For example, it is possible to chart the results of tossing dice that are weighted in different ways and to study the different shapes of the resulting graphs as well as the way in which the graphs approach the mathematically expected probabilities as the number of throws increases.

Here is an illustration of how this might work with two different dice:

FACE	I	II
1	1	1
2	2	2
3	3	2
4	4	2
5	5	3
6	6	3

Die I is the standard craps die. Each face has a different number and in any throw all the numbers from 1 to 6 have an equal chance of coming up. This means that on a given toss each number has a one in six chance of occurring — that is, the probability of its occurrence is 1/6.

Also, over an indefinite number of tosses each number should appear equally frequently. However, there are often runs of a particular number which skew the graph and it is interesting for students to discover how these runs of luck ultimately even out.

The first twelve tosses of a regular die might produce this skewed graph:

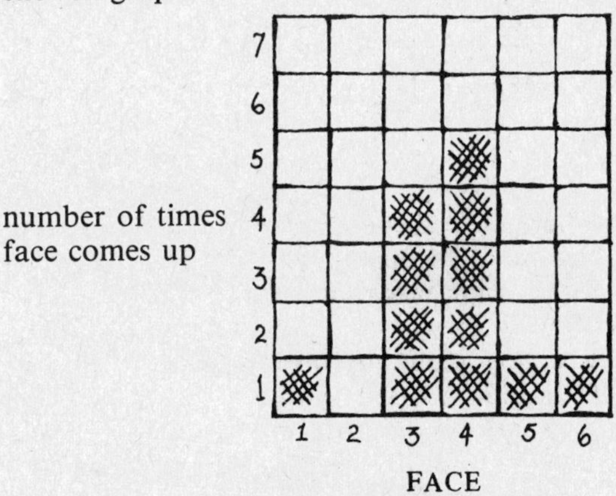

This is not at all what would have been predicted on the basis of the probability that each face would appear the same number of times. The mathematically predicted graph would look like this:

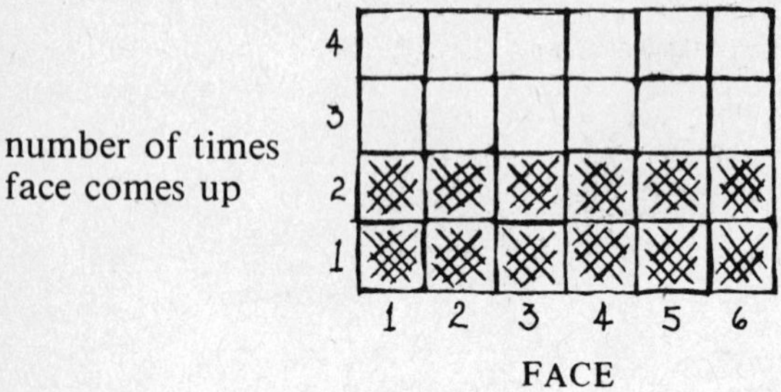

However, after thirty tosses of the die it is likely that things will even out more:

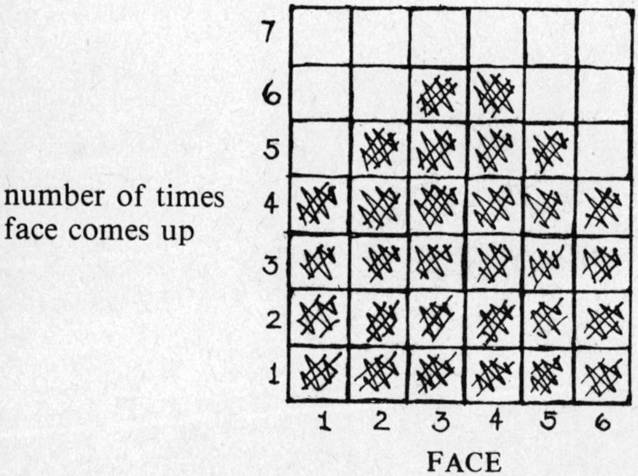

After several hundred tosses the graph of actual tosses will look very much like the mathematically predicted graph. The relationship between actual events and those predicted by a mathematical analysis of the probabilities involved can easily be discovered by young children through tossing dice or flicking a spinner. Adults might get bored by tossing a die a hundred or so times, but I've seen primary school children toss a die fifty times a day for two weeks and record the five hundred results on a graph with fascination. They are intrigued by seeing the laws of probability work out.

Sample tosses of the second die described above will provide different results because the weighting of the die changes the probabilities so that 1 has a probability of 1/6; 2 a probability of 3/6 or 1/2; and 3 a probability of 2/6 or 1/3. After thirty tosses the mathematically predicted graph is:

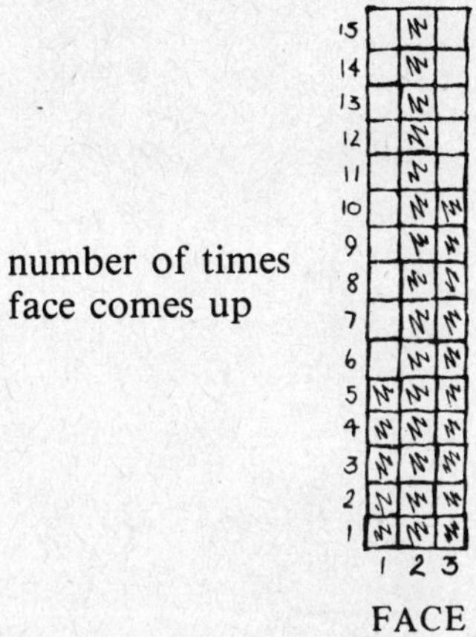

The actual results might be:

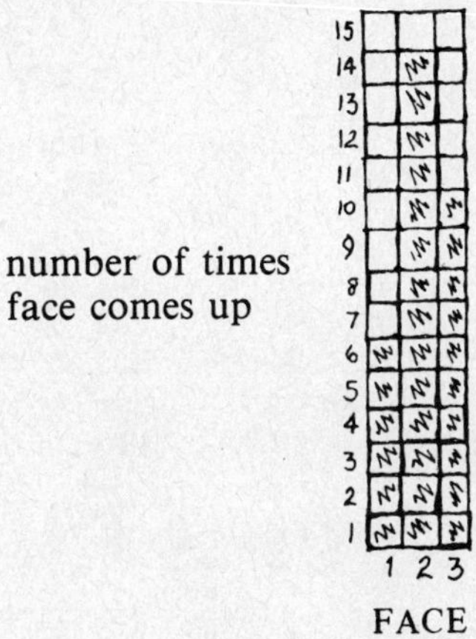

It would be interesting for students to make their own dice, predict and draw the mathematical probability graphs, and then see how long it takes for the graph of actual tosses to approximate the mathematical predictions.

Primary school children can be introduced to probability theory this way. At the same time fractions are introduced and used, you might introduce the subjects of probability, making and tossing dice, and modifying games by changing the decision-making procedures. Perhaps the use of dice with different coloured faces would be less confusing for young children than the use of numbers. Then a graph could be colour coded and therefore understood more easily — e.g.:

number of times
face comes up

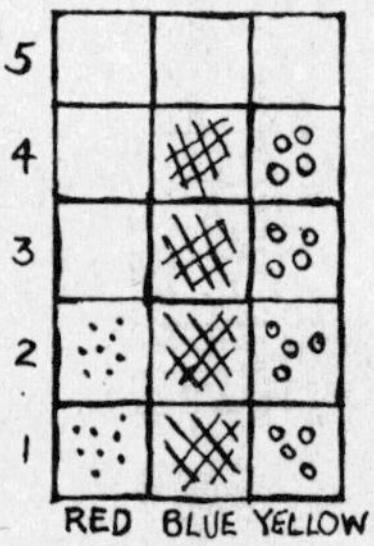

FACE

* * * *

There are other ways to label dice which are similar to the ways some of the spinners described before were labelled:

Face	Markings			
1	stop	kiss	double points	take a partner
2	go	hit	0	go it alone
3	stop	shake hands	triple points	join a family of three
4	go	leave	0	take a partner
5	stop	smile	single point	join a family of three
6	go	cry	0	
	stop—3/6 go—3/6	each—1/6	0—3/6 other—1/6 each	each—2/6

There are two other kinds of dice I want to describe briefly. One comes with a French game called The Voyage of Ulysses. Each player in the game commands a boat which starts at Troy and journeys over the board (which is a map of the Mediterranean Sea charting Ulysses' supposed voyage home after the Trojan War) to Ithaca. There are certain 'fate' points marked on the board. If a player, tossing an ordinary die, lands on one of these points (Ulysses' route is divided into small units of equal lengths marked by points — e.g., Ithaca 

Troy), he or she has to toss the die of the gods. This die is labelled:

1	Poseidon	4	Athena
2	Zeus	5	Aeolus (wind)
3	Hades	6	Hermes

Three of the gods — Zeus, Athena, and Hermes — are considered favourable to the voyage, and three — Hades, Aeolus, Poseidon — are considered opposed to the voyage. If a favourable god comes up, the player is given additional moves or points (as determined by a chart in the rules book); if an unfavourable god is one's fate there are punishments (such as backward moves, loss of points).

This die introduces the complex element of fate into the game and something like it could probably be introduced into many chance games to complicate the play. For example, in a simple car race game there can be fate squares and a die labelled:

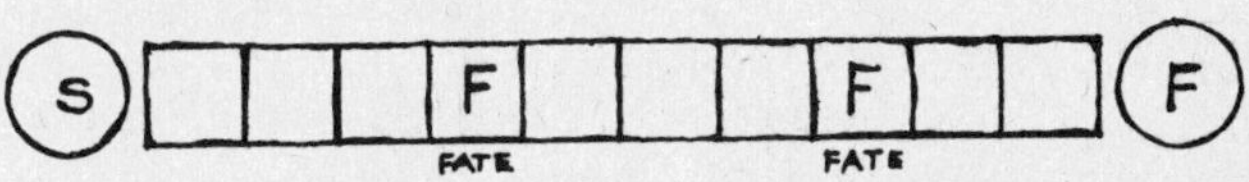

Face	*Label*	*Penalty or Reward*
1	crash	lose 2 turns
2	run out of gas	lose 1 turn
3	spectator on track	go back 2 squares
4	quick tyre change	gain 1 extra turn
5	extra gas	go ahead 2 squares
6	new engine works better than expected	gain 2 extra turns

Another interesting die comes with an English game called Exploitation. This die not only indicates how many squares a player can move but shows the directions of the move as well.

Dice which indicate the directions of moves as well as the moves themselves have not been explored very much so far as I can tell. There are many possibilities. For example, it might be possible to abstract the moves of chess pieces and try a new version of chess which consists of five pieces for each player, a five-by-five board, and a die indicating the six different pieces involved in chess:

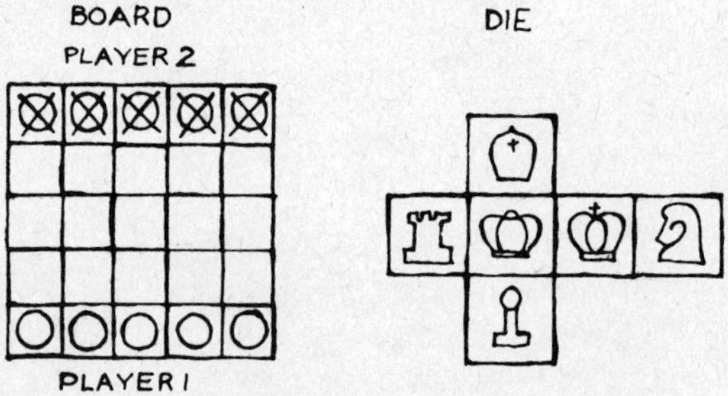

The goal of the game is to capture the other side's pieces. The moves are determined by tossing the die. Whatever piece comes up on the die determines the move that can be made on that turn. Let me clarify this by an example:

1. Player 1 tosses the die, comes up  and moves any one of his or her pieces the way a bishop moves in chess. Here is one possibility.

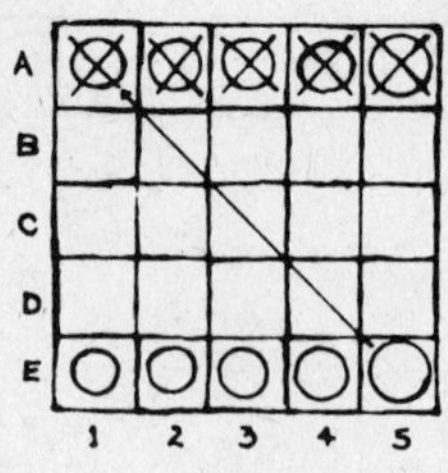

Player 1 moves the piece at E5 diagonally (as bishops move) to A1 and takes the piece on A1 as bishops take in chess. After this move the board looks like this:

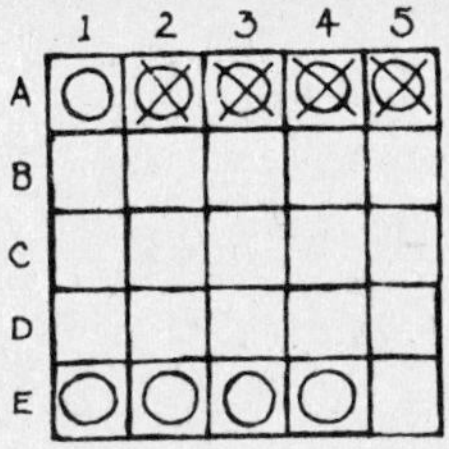

2. Player 2 tosses the die and comes up with  and moves the piece on A4 to C3.

This game might be fun for people beginning to play chess, since the board is very simple and strategy can only be developed a move at a time. The constant shifts in power of the game introduce chance into the strategy game of chess and at least for me provide an opportunity for learning how to think and plan quickly and flexibly. In chess the strategy evolves slowly and consistently. In this game the strategic advantage often changes suddenly and no long-range plans can be made because a toss of the die can wipe them out.

* * * *

After playing around with dice and spinners for a while it becomes clear that dice are identical in structure to spinners with six equal parts. They are both physical models of a random selection process with six equally possible outcomes. Young people discover this structural identity quickly and generalize the discovery to other instances. For example, a bag of six different coloured marbles from which one is selected per turn at random is also identical to a die or six-sectioned spinner. So is a hat from which are selected at random one of six folded pieces of paper with different marks on them.

The discovery that similar mathematical ideas can have different models, as well as the notion of structural identity, are both central to an understanding of the development of mathematical systems and models of systems. They are also notions that are very difficult to communicate to young people in an abstract or verbal way. Using games or game components which embody these ideas provides the student with experiences to draw on in order to understand some of the sophisticated ideas of mathematics.

There is one more use of dice I'd like to mention. Using two dice, one labelled alphabetically and the other numerically, it is possible to cover all the points on a five-by-five grid:

Face	1	2
1	1	A
2	2	B
3	3	C
4	4	D
5	5	E
6	6	F

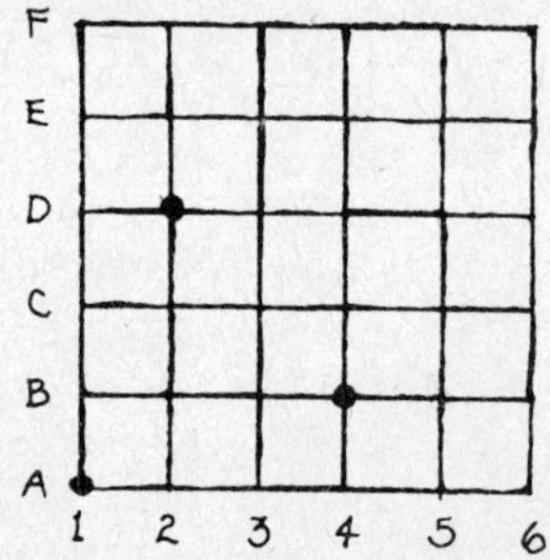

Thus tossing the dice three times, (A,1), (B,4), (D,2), would indicate three points on the grid. It is then possible to create a chance version of battleship on a five-by-five board with the coordinates called out determined by the dice rather than the players' guesses. (See Section II of this chapter, on boards, for instructions on how to play battleship, then try to introduce these dice into the game.)

A different version of the same idea can lead to a chance version of noughts-and-crosses. This time a three-by-three grid is used so the dice can be relabelled:

Face	1	2
1	1	A
2	1	A
3	2	B
4	2	B
5	3	C
6	3	C

and the squares on the noughts-and-crosses board labelled:

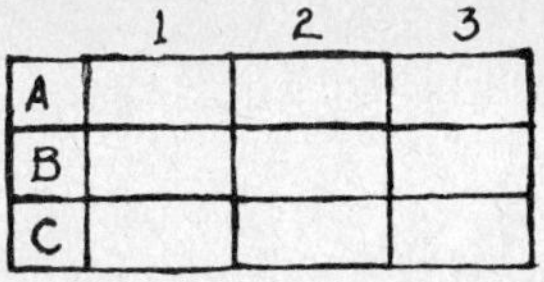

One other rule will be needed: if on a throw of the dice a square already occupied comes up, the player casts again until an empty square is indicated. Here is a sample game:

Player 1 (X): (2, A)
Player 2 (O): (3, B)
Player 1: (3, B) occupied,
  throws again (1, A)
Player 2: (2, B)
Player 1: (3, A) wins

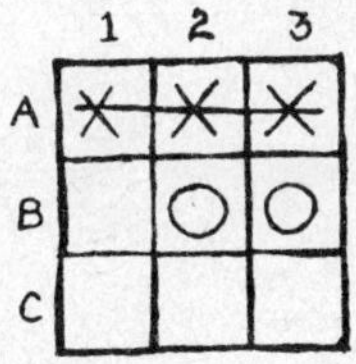

* * * *

Other chance devices such as card decks or pieces used for drawing lots are not structurally different from dice or spinners. In fact one can look at a deck of cards as a spinner with fifty-two divisions or as a die with fifty-two sides.

However, the names of the cards in a usual card deck add a further complication since each card's name has three components — the value (ace, 2-10, jack, queen, king); the suit (spades, diamonds, clubs, hearts); and the colour (black or red). Thus the deck can be sorted and ordered according to:

1. individual cards, with each card having a 1 in 52 chance of appearing in a random drawing:
2. value, with each value (represented by four cards) having a 4 in 52 chance of being drawn;
3. suit, with each suit of thirteen cards having a 13 in 52 chance;
4. colour, with each colour having a 26 in 52 chance of being drawn.

Because of these different ways of ordering a single deck dozens of games have been generated.* When used as

*For card game rules see the section on card games in the bibliography at the end of this chapter.

decision devices in other games, cards can be used to decide such things as who goes first, who wins a tie, who chooses which piece to play, etc.

Many games, however, do not use the standard deck of playing cards but create their own decks. In Monopoly, for example, dice are used to determine the number of squares to be moved, and a deck of bonus and penalty cards is used to introduce a random element into the play. Thus if a piece lands on one of several special squares the player picks from the deck and will be instructed to pay a fine, collect an award, go to jail, etc.

A deck of penalty and bonus cards, especially one created by students, can add a great deal to a simple chance-determined race game. There are a number of ways this can be done. Given the following board,

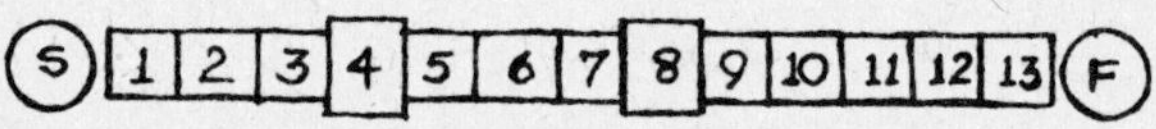

the player who lands on square 4 or 8 has to pick a bonus/penalty card.

The simplest bonus/penalty deck merely indicates moves:

1. Go back two squares
2. Go forward three squares
3. Take an extra turn
4. Lose your next turn.

Games with specific themes naturally ought to have cards that relate to these themes. For example, in the bee game discussed in Section II of this chapter, a few spaces could be designated as special and the cards could say such things as:

1. Go directly to a flower and collect its pollen.
2. You spilled your honey, return to your hive and begin again.

In chance games it is best to make picking a card mandatory — if you land on the square you must pick a card. In strategy games picking a card can be made into a gamble. For example, in checkers one square can be called special. Whoever lands there can pick a card if they wish. Some cards might say:

1. Make kings of any two of your pieces you choose.
2. You can take any of your opponent's kings you choose.

3. You lose your piece nearest this square (in case of tie your opponent chooses which piece you lose).
4. You lose your next piece to be made a king.

In this context taking a card or refusing to draw one is a calculated gamble. You might win a lot or lose a lot, or by not drawing leave the game as it is. Of course, if you're losing, the gamble might make more sense than if you're already ahead.

Another form of penalty/bonus deck might consist of question cards. If the questions are answered correctly you gain some moves, if not you lose. For example, questions like these might be on the cards:

1. What is [(2 + 4) — 7] x 2?
2. With your eyes closed, can you name all the students in the class?
3. How many classes are in this school?
4. How old is your teacher?
5. Can you name six different species of birds?

* * * *

There is no set formula for the proper balance of chance and strategy or suspense and predictability that ought to be included in a particular game. As you make a game with your students, play it together, change it and then play it again as you discover whether what you are doing is of interest or value to others.

5 Setting goals

Most games are competitive and have clearly defined winners and losers. In fact the need to maintain a competitive edge sometimes leads to absurd and self-defeating behaviour. Last year I held a small class on games for second and third graders. My goal was to get the students to teach each other traditional games as well as to develop new games of their own. One third-grade boy knew how to play chess. None of the other children knew the game and so I suggested he teach one other person how to play. He refused at first — he knew chess and the others didn't; that was his advantage and he wasn't about to give it up. Let them go and learn it themselves if they wanted to. But after a few weeks he reluctantly agreed to teach the game to one of the girls in the class. He set up the chess board and began a game, telling her which moves to make and then moving his own pieces. The first game he checkmated her in three moves; he kept giving her stupid moves so he could humiliate her and guarantee he would win all the time. After a half hour the girl didn't want to learn the game. I came over to them and began to teach her myself, to get her to feel confident about her ability to master the game.

Afterward I spoke to the boy about how he was going about teaching. In his mind winning was the central fact of the game — playing was merely a series of tasks one performed in order to win. There was no joy in the beauty of the game or pleasure from merely playing.

Teachers have to be careful how games are taken by their students. The simplest game can turn into a nasty confrontation. Students can mock losers, drive away less skilful players, come to look upon games as unpleasant teacher-imposed tasks.

I have found that de-emphasizing the winning-losing aspects of games by creating new games or changing the goals or rules of old games, or by looking at games as interesting models to study, helps the students to respond with curiosity rather than competitiveness. They begin to comment on well-played games, analyse mistakes, look upon losing as a natural thing to happen in learning something new. They change things around just to see what will happen and don't look upon rules as holy and unalterable. In other words they play and don't merely compete.

Some liberal teachers believe that all competitiveness should be removed from the classroom; that only co-operative games should be allowed. I don't feel that way. There is nothing wrong with occasionally testing yourself against someone else. The more crucial thing is your attitude towards the outcome of the game. In some cultures it is considered the winner's obligation to avoid humiliating the loser. In others, games are abandoned before they are finished so that a winner does not have to be publicly declared even though everyone knows who probably would have won.

It is possible for two people to respect each other's skills, to test themselves, and then to finish with the game and not consider the outcome a judgment on the moral or intellectual capacities of the winner and loser.

The same problem of whether competitiveness is at all desirable has come up in athletics recently. Some critics of the Vince Lombardi attitude ('winning is not just part of football, it is the only thing') have gone to the other extreme and suggested that all competitive athletics be eliminated. Others like Jack Scott* have suggested that one should try to change people's attitudes toward winning rather than eliminate athletic contests altogether. It is possible for the winner of a close race to feel grateful to an opponent for providing the occasion to test him or her, and for the loser to feel part of a collective effort to become competent. It is possible, in other words, for people to compete in a communal spirit rather than an individualistic one. I think the same thing is true of people playing games.

There are a number of ways that winners are determined in competitive games. Some of them which I will consider in some detail are:

1. by points;
2. by getting somewhere first;
3. by capturing or trapping one or all of the opponent's pieces;
4. by winning a majority of games over a period of time (three out of five, four out of seven, etc.).

1. Winning by Points

There are several ways to set number goals. The first person to accumulate a certain number of points (like fifty or a hundred) might be declared winner, or the person who has the highest number of points after a certain number of plays (like innings, quarters, hands at cards). This can also be reversed so that the last person to get a certain number of points or the person with the lowest score wins.

Sometimes changing the scoring system drastically changes the nature of play. Think of the following possibilities:

1. In football a touchdown is six points, a field goal three

**The Revolution in Athletics* (Free Press).

points, a safety two points, and one point can be made after the touchdown. Imagine the way game strategy might be changed with the following scoring systems:

touchdown	3	3	2
field goal	3	2	6
safety	2	4	3
point after touchdown	1	1	4

In my class this year I've experimented with scoring systems on the playground. One day we play basketball with each field goal and foul shot counting one point. Another day the foul shot is three points and the field goal one. This change causes, of course, fewer fouls; but it also encourages students to think and talk in class about the ethics that are built into the game, and how scoring affects it.

We do the same thing with kickball, football and baseball.

2. In the number dice game described on page 148, the winning score can be changed from the lowest to the highest, leaving the rest of the game unchanged.

3. A series of simple two-colour dice games can be developed with a variety of goals:

Score — number on red die minus number on white.

Goal — (a) first player to score exactly ten points (using plus and minus numbers) wins; or
(b) last player to score ten points wins; or
(c) first player to score minus ten points wins;

or

(d) last player to score minus ten points wins.

4. I once noticed some students shooting craps in the most puzzling way. On one roll of the dice a six and a three came up. The player shooting called out five. Then the next player rolled a one and a three and called out ten. The first player rolled again, got a five and a four, and called out five. I thought the kids were putting me on and they laughed as I observed them, dumbfounded. When a six and a three came up a second time the player called out five again. There was clearly some regularity to the game and finally I asked to be

taught how to play. It turned out to be very simple. The students had changed the rules of craps so that your score was the sum of the numbers opposite the ones that came up. In other words, the faces touching the floor gave the score. This introduced an extra complexity into the game. You had to memorize all the opposite faces on a die — one opposite six; two opposite five; and four opposite three — and then call out your score without seeing it. It wasn't that difficult, since they soon discovered that the sum of the faces opposite one another always added up to seven. Therefore, if a five was rolled you could quickly compute, 7 — 5 = 2, to get the opposite face. Thus, with two dice, six and three showing is transposed into: 7 — 6 = 1; 7 — 3 = 4; and 4 + 1 = 5. This neat mathematical reckoning was performed with ease by students who refused to solve the simplest arithmetic problem in a formal classroom setting.

An interesting exercise to use in class is to find out about the scoring systems of the following games and modify them: bowling, pool, billards, bridge, dominoes, craps, roulette. What changes in strategy do the modifications make? Do the games become more or less interesting?

2. Getting Somewhere First

In race games the winner is usually the first player whose piece crosses the finishing line. There are other ways of finishing first:

1. in card or matching games, getting the right number of pairs or triplets and going out first;

2. being first to pass a certain number of points (as described previously on page 175);

3. being first to achieve a certain criteria (such as three Xs or Os in a row, as in noughts-and-crosses).

These criteria can usually be reversed so that the last person to finish wins. It is possible, for example, to change the rules of noughts-and-crosses so that the person who is forced to get three marks in a row *loses*. This leads to a whole new strategy for playing the game. I've tried reverse noughts-and-crosses and find it more interesting than the usual version.

3. Capturing and Trapping

The goal of checkers is to capture all the opponent's pieces, whereas the goal of chess is to trap the opponent's king.

Capturing and trapping have many variants.

For example, in the version of checkers where jumps are required, the rules could be changed so that:

(a) as soon as a piece reaches the eighth rank it is taken off the board;

(b) the winner is the player who has captured the *fewest* number of pieces.

The game is complex because jumps are mandatory and it is often possible to move into a position where your opponent is forced to jump you.

There are a number of themes that come to mind for creating possible capture games:

— Wolves vs. rabbits and other games involving the balance of nature: If the wolves capture all the rabbits or reproduce too vigorously they end up starving or destroying each other. Captures are made when a wolf lands on a square occupied by a rabbit. If there are too few rabbits left (some minimum number can be determined for the game), the wolf is required to capture any other wolf within a three-square radius of where its move ends.

— Big fish eat little fish.

— Spider and fly games.

— Hunter vs. animal.

— Nation vs. nation. (Many war games already exist, some of which are very sophisticated and require considerable calculation, strategy and time.*)

4. Winning a Majority of Games

Many games are played again and again, with the winner being determined after a series of games has taken place. The World Series in baseball is won by four victories out of a possible seven games. In chess tournaments, anywhere from ten to over twenty games are played with one point gained for a win, half a point for a draw and zero for a loss. The winner is the player with the highest score at the end of the tournament, and ties are possible. In the world championship match over twenty games are played before the champion is decided upon. No single game is considered definitive enough to decide upon who is the finest player. If infinitely long series

*See the bibliography at the end of this chapter for comments on what are euphemistically called 'strategy and conflict' games.

of games were held, there would be no need to declare ultimate winners and people could attend a lot more to the quality of play and the beauty of the game itself rather than to the idea that for every human activity there has to be a champion.

* * * *

Not all games are competitive, though many people find noncompetitive activities of all kinds boring. My daughter Erica refuses to play winning games and has invented a series of ingenious cooperative games, one of which is played with some of the blocks Jean Doak made (see page 148 for a description) as well as a spinner labelled in the following way:

We used six each of the arrows, squares, triangles, circles and pentagons. The goal was to build an interesting design on the floor. We flicked the spinner and the person whose name came up could pick a piece and add it to the design. We played the game again and again, adding new pieces, creating different designs each time. When I explained the game to a friend of mine, he said it seemed boring and pointless, especially since nobody won anything. Yet many children can still enjoy watching a design evolve for its own sake, without the rewards or punishments of competitive games.

This simple building game can be modified to form a variety of *collaborative collage games*:

Instruments: glue, staples, scissors, hammer and nails.

Decision makers: spinners or just the simple order of seating in a circle.

Pieces: shells, beads, marbles, newspapers, magazines, blank paper, stones, pieces of wood — all either distributed among the players or put in the centre of the playing space.

Board: a large piece of butcher paper or quarter-inch plywood.

Moves: People take turns building a collage, cutting or pasting, choosing something from the centre or from their own pile, depending on how the game is set up.

Goal: The goal is to complete a collective work in a certain number of turns, or to use up the materials, etc. One can make the game more complex by setting a theme for each collage so that there is some idea or statement that everyone is moving towards: e.g., sex/sexism; oppression/race; turning things upside down; taking a gamble; living dangerously; opposites; conflict/violence.

For example, I recently did a collective collage/game/poem with sixteen teachers. Everyone had a three-by-five index card on which they wrote one word — any word which came to their mind. Then we went around in a circle, laying the cards out on the floor and building a poem. The words people chose were: today, earth, home, love, joyous, heart, Jon, above, aches, sun, please, golden, gentle, why, whistle, people.

The first word put down was: please

The next person added: today please

Then we got: today please earth

and: today please above earth*

and: today please
sun above earth

and after everyone had contributed:

why today
please whistle joyous heart
people home love Jon
gentle golden sun above earth aches

We then took the same words and built up a new poem.

Building collectively can be delightful, though for many people it seems pointless. Why develop a collage or build a model or put together a poem? I can only say that it teaches me about other people and myself and provides an opportunity to think and play without being judged, rewarded or punished.

There are some collective building games which are similar to these collage games. It is possible to set themes for collaborative games such as:

— making a city (a) that pollutes itself to extinction; (b) that eliminates pollution

*We agreed that words could be inserted within the original 'poem' so long as the initial order of the words was not changed.

— developing a new culture
— settling uninhabited territory
— developing:
  utopias
  limbo
  a city for people
  a city for commerce
  a city for the rich
  a city of the young
  an island that can survive
  a self-contained community

Developing these games is more complex than merely creating a collage out of available materials. Constraints and principles of development have to be articulated. The resources available to the players have to be defined and the moves of the players and the consequences of these moves must be spelt out. Let me give an example:

— The goal of the game is to create a town for 6,000 people.

— There are four players — one whose role is to lay out the streets, a second who creates the housing for the people, a third who is responsible for jobs to keep them alive, and a fourth whose responsibility is recreation and pleasure.

— Each player is given materials appropriate to his particular task:

for player 1, a map of the terrain on which the town is to be built;
for player 2, a series of descriptions of housing possibilities (pictures of private homes, projects, apartment buildings, etc);
for player 3, a list of industrial and business possibilities, ranging from large industry to small shops;
for player 4, a series of descriptions of parks, playgrounds, night clubs, restaurants, malls, etc.

In addition to what has been listed, a number of goals for the community have to be established, such as:

— full employment or maximum benefit for the rich;

— individual property or communal ownership of the town;

— maximum ability to grow in the future, or an insistence of developing a self-contained and stable community;

— a dependent relationship on surrounding communities or independence.

The goals have to be pre-set. They can either be decided on by the players or pre-set but changed each time the game is played. Then each player can develop his or her own component independently of the others and later everyone can come together to negotiate and put down a final plan. In the last step, some people who have not been involved in the planning would analyse the results, to see if the original criteria have been met.

One may wonder whether this project is a game at all. But, like other games, it can be done over and over again, it involves invention and play, there is an outcome, and the players can modify the rules and learn from the activity without being committed to anything beyond the activity itself. The boundary between playing games and 'serious activity' is not clear. Play and work are not really distinct — especially creative work and creative play. One plays to explore possibilities and then works to realize what seem to be the best ones. To play without competing might help us learn to work without competing, or doing so in vicious ways.

6 Using games in the classroom

Games can be used throughout the curriculum and they can become a separate part of learning as well. In the K-2 grade class I will be working with next year, we have a number of learning centres. Each centre is devoted to a particular subject or to the exploration of a theme. Thus there is a reading centre, a maths centre, a shop, a kitchen, a theatre-fantasy centre, as well as a centre devoted to the study of living things and another one devoted to playing and making games.

Reading, writing, maths and games will be concentrated in their specific centres, but they will also be worked into the other centres. For example, in the theatre-fantasy section, students will be making costumes, planning and building sets, writing and performing plays and dances. In the maths centre there will be maths games. There will also be a number of specific projects planned. The students will, for example, have the opportunity to plan and build several domes and will be encouraged to compile a handbook so that other primary school children can learn to build domes too.

In the life forms centre, games on ecological relationships, animal matching and identification games, as well as dinosaur and bee games will be included. Students will also have the opportunity to create new games. The game centre will be devoted to strategy games (like chess, Chinese checkers, minimal chess, checkers, Go, NIM, Wari, etc.) as well as to many chance games that were made by students last year. Our emphasis will be on traditional games on the one hand and student-generated games on the other. We'll avoid commercial and expensively packaged games as much as possible.

Since emphasis will be put on game making, a part of the game centre will be a workshop. In this game shop will be the materials needed to make games — boards, pieces, decision devices, as well as the pens, glue, paper and other tools students might need. Of course the shop itself will be used for the construction of more elaborate things such as carved Wari boards, or hand-made chess pieces.

To begin with, the following materials will be available in the game centre (we'll modify or add to this list as our needs develop):

Boards:

1. checker boards.
2. three-by-three boards mimeographed onto thick paper.

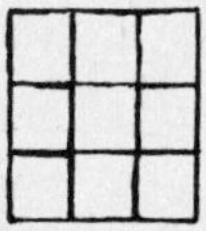

3. four-by-four; five-by-five; six-by-six; seven-by-seven boards as well.
4. graph paper so that students can make their own boards.
5. a variety of race-type boards plus cardboard, rulers, etc. so students can make their own race boards.

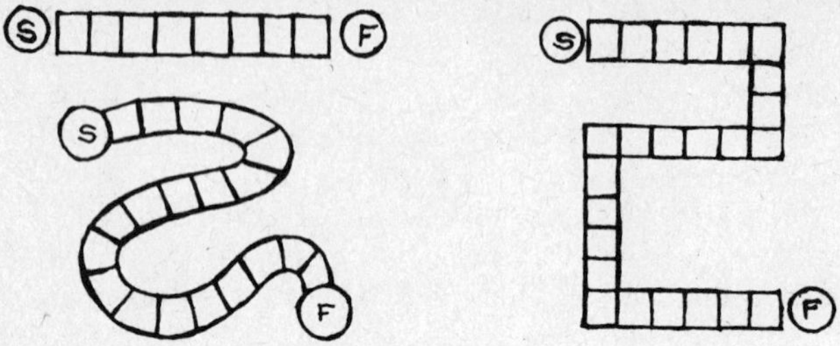

6. a sample Wari board and a collection of egg cartons and paper cups.

7. a box of ceramic floor tiles of different colors so that students can experiment with fitting their own boards together. The tiles will be both square and octagonal.

8. a Go board.

Pieces:

1. a large box of plastic animals and people.
2. lots of poker chips.
3. a box of plastic or metal cars and trucks.
4. several sets of chess pieces (which can be broken up so students can make new chess-type games).
5. several boxes of stones and nuts for use with Wari-type games.
6. boxes of black and white Go stones.
7. three-by-five index cards and lots of magazines and photographs that can be used to decorate game boards or make cards for matching games.
8. empty boxes for the students to fill with objects that they find or bring to school that can be worked into games.

Decision-making devices:

1. a box of white dice.
2. a box of green dice.
3. a box of red dice.
4. if possible a box of blank dice and marking pens. If these dice can't be obtained we'll use self-sticking labels to cover the faces of ordinary dice, or cubes that are used in some math games.

5. a range of different-coloured marbles which can be used in lottery drawings and as pieces for Chinese checkers or other marble games.

6. spinners, most with blank faces, some with yes/no or numbers already written on them. Also there will be lots of cardboard arrows, a hole punch, brass paper fasteners, compasses, cardboard, so students can make their own spinners.

7. letters and numbers written on cards, [1], [2], etc., to be used for lottery drawings and also for pieces in maths and reading games.

8. several decks of cards, and blank cards for people to make up their own decks.

In addition to this material, which will be stored on shelves in a bookcase that is serving as a partition to mark off the game centre, there will naturally be a number of games available. The best way I've discovered to store games is by using heavy see-through plastic bags with plastic zippers on top. Commercial games come in boxes that are usually three-quarters empty and take up a lot of space. The plastic bags are very compact, the students can see the parts involved in the game, and instructions can be included. The bags themselves can be hung on the inside of a cabinet door or on the side of a bookcase. It is possible to store nine or ten games efficiently on the side of a six-foot high bookcase in this way:

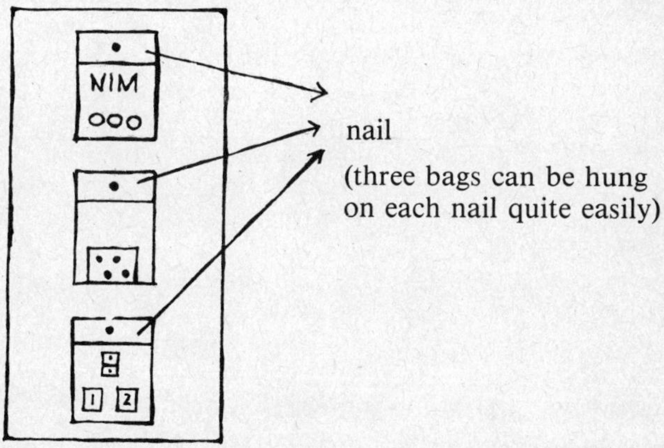

Large playing boards can be stored separately on a shelf in

the bookcase and notes indicating which board, numbered on the back, goes with which game can be left in the plastic bags containing the pieces.

Games and game parts have to be stored carefully. Otherwise students will be overwhelmed by the number of different things available and will end up unable to maintain them. By showing clearly where things belong, and making it as easy as possible for the students to keep everything in order, a teacher can help the students to function with ease and free them to create.

I have known some teachers who believe that open education implies that students can do anything they want with the materials available to them. They let the students demolish the classroom and after a few weeks there is no comfortable space to work in and no usable material left. I believe that students can only be free to choose and learn if they accept these four minimal constraints:

1. no one is free to harm another person;
2. no one is free to destroy the work of another person;
3. no one is free to prevent another person from doing work;
4. the maintenance of the environment is a collective responsibility.

This doesn't mean that the space organized by the teacher is an absolute that cannot be changed or that the students shouldn't have a major part in planning. But if people are to survive collectively and work creatively they have to develop and maintain a comfortable place to work where tools, materials and other resources are available when needed.

In our game centre we hope to develop as many game playing surfaces as possible — table tops, the floor, small round tables made out of old wire spools obtained from the telephone company. Tentatively the centre will look like this at the beginning of the school year (of course it will change quite a lot as it is used):

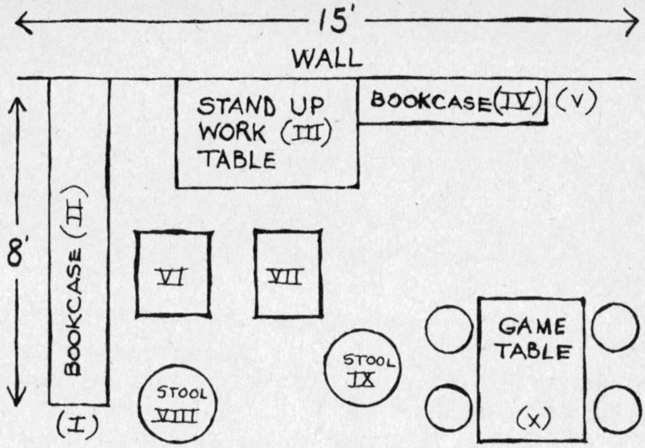

I — side of bookcase used to store games
II — bookcase to store game parts
III — stand-up table for game making
IV — bookcase to store game boards, other games to be played.
V — side of bookcase to store more games
VI and VII — two pillows (there will be more). Can be used to lean on and games can be played on floor.
VIII and IX — table spools for game-playing. These are cut low enough so that students can kneel or sit on pillows.
X — table and four chairs for game-playing.

This centre could accommodate, when things are going smoothly, between ten and fifteen primary school children. Though our class will consist of five-, six- and seven-year-olds, such a centre could be of interest to students throughout their school years.

I hope to introduce the students in our class to the centre gradually over the first few weeks. It has been my experience that it is easy to overwhelm and confuse children (especially if they have had experience in traditional classrooms) if they are not shown some of the possibilities provided by their environment. My present idea is to sit down with all the students and show them a simple game (perhaps a matching or lotto-type game, minimal checkers or an easy race game) on the first day of school. Then whoever wants to play the game can do so. That first morning, the students will also be introduced to the other centres, and they will be able to choose a particular place to explore and a particular activity they can begin there. They will also be given the chance to explore at random, but I feel it is important to begin by providing a focus for their initial explorations of the classroom.

From the very beginning we hope to make it clear to the students that it is our collective responsibility to put things away and to respect each other's work. In my experience, preaching or using punishment has rarely developed a sense of responsibility in children. The need to maintain the environment is primarily a question of group survival, coherence and comfort, and should be dealt with in a very concrete and unambiguous way. If kids naturally take care of games, that's fine. If someone leaves something around, he or she has to be shown where it belongs. Sometimes very young children take and use more things than they can reasonably be expected to put away. Then it's up to the adults to help them put things back together and to enlist other children as well.

I don't like to dwell on the question of house-keeping, but in the last few years I've witnessed too many examples of schools and classrooms that have become depressing, filthy places in the name of open education. I believe that if adults don't care and if they let children do whatever they want the children feel abandoned and rejected, and behave as if they hated both the adults who pretend to give them freedom and the place in which they are supposed to be free.

I anticipate that after the first few days of school some students will have discovered a lot more about games and the game centre than others have. There will be a few who want to learn new games, a few who will be set off by the mere suggestion that they can create their own games. The students will begin to teach each other, and I see myself playing a smaller and smaller role as the year goes on. I will show a few students new games and suggest that they teach other students. Also I intend to make up some new games myself since I enjoy doing this and have found that the more interested an adult is in an activity the more children will try it themselves. Finally, I'll play games with the students if they ask me and leave them to play with each other when they want to.

* * * *

But why teach games anyway? What do the kids learn and how do you evaluate it?

Let me try a somewhat indirect answer to these questions by considering one of the clichés of contemporary education: begin where the students are; start your programme with what they know.

Beginning where the students are is usually interpreted by teachers as learning, through various 'diagnostic instruments', how much of the standard curriculum each student has mastered, and then starting them at their own levels. This is referred to as 'individualized instruction' and is little more than channelling students through the traditional curriculum at different rates. New words like 'individualized instruction' generate new programmes, or rather the repackaging of old programmes.

For example, IBM under the trademark of SRA sells individualized learning kits which are little more than traditional reading materials organized into a system by which each child works alone from level to level rather than with the rest of the class. In this setting the teacher's role shifts from conductor of the whole class to systems manager. The manager has to be sure that each person enters the system at his or her appropriate level and proceeds through it in an orderly manner.

But there is a completely different way of interpreting where the students are which focuses on what the students bring with them to school, what they learn at home or on the streets, from the adults on television, or in the playground. Most students who are failures in school, who, to judge by the results of standardized tests and teacher's reports, appear stupid, know all kinds of things that are never considered important in school or used as the foundation upon which to learn new things. Over the last few years I've heard teachers, and have occasionally caught myself, saying things like:

'Put that comic book away. It's time to do reading.'

'Stop playing cards. It's maths period.'

'Shut off your transistor. It's interfering with the choral-music lesson.'

In other words:
'Stop reading. It's time to study reading.'
'Stop using maths. It's time to study maths.'
'Stop listening to music. It's time to study music.'

Beginning where the students are implies knowing your students, not just as they appear in school, but how they are during most of their day. Teachers, especially those who work in communities they did not grow up in and do not live in, generally don't know enough about their students. It is necessary to spend time in the community, to learn what parents teach their children, to observe the games the adults or the adolescents play, to know something about gambling, about the verbal games people play, about jump rope games and street rhymes — in other words to know the culture not as an anthropologist from an outside community but as a participant and celebrant.

In learning who the student is, what the community is like, one comes upon games everywhere. People throughout the world learn through play, practise through play and relax through play. By beginning the learning process in school with the games of the community, a teacher permits the students to remain in contact with a familiar world while they reach out to more abstract issues. Playing a game can lead to analysing the game, to research embodied in the game, to historical investigations about the origins of the game, to modifying the game, etc. I don't mean to imply that games are the only, or even the central, aspect of people's lives. However they do provide a good starting place.

Recently I've explored the mathematical skills and knowledge that play a part in everyday life and translated them into learning experiences that can extend into the classroom and then back to the community.

I've developed a list of different areas of knowledge that can be used as a guide to studying what mathematics the students have and therefore to generating starting points for units in math. Not surprisingly, games can be found in almost all of these.

Mathematics are embodied in everyday life through:

1. *The way people locate themselves and others in space,* through the territory they claim both within their homes and in the community; through the boundaries they draw and

those they respect; through the games of hide-and-seek and chase they play; through the places they gather to play cards or dominoes or to meet members of the opposite sex or to take babies out for a stroll, play ball, go to the movies, a bar or a restaurant. People also define themselves by whether the territory is foreign or dangerous. Knowing some of this information, and using it in the classroom, can lead to the study of maps, boundaries and boundary games, topology, measurement of areas, making floor plans, replanning a community, etc.

2. *People moving around*, getting from one place to another by foot or train, bus, car or plane. Different modes of transportation are appropriate to different forms of mobility. It doesn't make sense to fly to your friend's house next door, or to walk 3,000 miles if you can fly. What is the shortest way to get from one place to another in the
country, in a game? What is the
kest way always the shortest? How
? How about animals — dogs, cats,
hines? Is speed always a virtue?
arise naturally from considering

others in time. Work hours, the
time, birthdays, anniversaries,
by the sun, moon, and stars,
celestial movement on human
f the clock, the internal clock,
rsonal rhythms, rhythm in
measurement of rhythm and
nts, non-Western calendars,
xplore all these social and
with a study of how people
get up, eat, go to sleep, work
ne and when they take their

w many relatives do people
back can they project? How
ns, aunts, uncles can they name?
How ... nches of the family can be traced?
How can family trees be diagrammed, relationships

represented abstractly? How are other relationships (in logic) represented? Which relationships are symmetrical (I am the cousin of my cousin), nonsymmetrical (I am not my father's father), transitive? (If I am a relative of X, and X is a relative of Y, am I Y's relative? This of course depends upon how my culture defines relatives.) Is it possible to make up family games? To study family relationships in other cultures or the range of family structures in our culture? What are the variations of kinship structure throughout the world? Can they be represented algebraically?

5. *Sharing*. What resources in the community are shared? What is the child's share, the mother's, father's, relative's, friend's? How are things shared or budgeted? What fraction (or percentage) of the resources available are used in what way? How do teams share responsibilities during a game? How do young children learn (or not learn) to share? What is the devil's share? How are our natural resources shared? How are cards or dominoes dealt out so everyone gets the same number at the beginning of a game? How are food, clothing, sleeping space, living space shared?

6. *Accumulating*. What is saved or hoarded instead of shared? How is it done, how are resources counted, what are they weighed or measured against? What is the value of a dollar? How are points accumulated in a game? How do you add or multiply large sums? How do you know you're not being cheated?

7. *Exchanging*. How are things traded? How is value set? How many dolls are worth one bike? How are trading cards accumulated or exchanged? How do you weigh goods, balance them against one another? What systems of weights and measures are used throughout the world? Is it possible to get along without money as a medium of exchange? Is there anything that is so valuable that you would never exchange it?

8. *Gambling*. How is fate dealt with? How are odds set? Why gamble? What is there to lose or gain? What gambling strategies do people use? Do you know anybody who has won? How much? What are the chances of winning? What do people feel about winners and losers? How are the winnings

those they respect; through the games of hide-and-seek and chase they play; through the places they gather to play cards or dominoes or to meet members of the opposite sex or to take babies out for a stroll, play ball, go to the movies, a bar or a restaurant. People also define themselves by whether the territory is foreign or dangerous. Knowing some of this information, and using it in the classroom, can lead to the study of maps, boundaries and boundary games, topology, measurement of areas, making floor plans, replanning a community, etc.

2. *People moving around*, getting from one place to another by foot or train, bus, car or plane. Different modes of transportation are appropriate to different forms of mobility. It doesn't make sense to fly to your friend's house next door, or to walk 3,000 miles if you can fly. What is the shortest way to get from one place to another in the neighbourhood, in the country, in a game? What is the quickest way? Is the quickest way always the shortest? How far can a person walk, run? How about animals — dogs, cats, pigeons? How about machines? Is speed always a virtue? These and other questions arise naturally from considering how people get around.

3. *Locating oneself and others in time.* Work hours, the moral issue of being on time, birthdays, anniversaries, holidays, time as indicated by the sun, moon, and stars, astrology and the effect of celestial movement on human personality, the development of the clock, the internal clock, twenty-four-hour rhythms, personal rhythms, rhythm in music, in life, biological rhythm, measurement of rhythm and time, winning, racing, timing events, non-Western calendars, white man's time — one can explore all these social and mathematical questions, starting with a study of how people divide up the day — when they get up, eat, go to sleep, work — when they get to places on time and when they take their time.

4. *Family relationships*. How many relatives do people have? How many generations back can they project? How many cousins, second cousins, aunts, uncles can they name? How many different branches of the family can be traced? How can family trees be diagrammed, relationships

represented abstractly? How are other relationships (in logic) represented? Which relationships are symmetrical (I am the cousin of my cousin), nonsymmetrical (I am not my father's father), transitive? (If I am a relative of X, and X is a relative of Y, am I Y's relative? This of course depends upon how my culture defines relatives.) Is it possible to make up family games? To study family relationships in other cultures or the range of family structures in our culture? What are the variations of kinship structure throughout the world? Can they be represented algebraically?

5. *Sharing*. What resources in the community are shared? What is the child's share, the mother's, father's, relative's, friend's? How are things shared or budgeted? What fraction (or percentage) of the resources available are used in what way? How do teams share responsibilities during a game? How do young children learn (or not learn) to share? What is the devil's share? How are our natural resources shared? How are cards or dominoes dealt out so everyone gets the same number at the beginning of a game? How are food, clothing, sleeping space, living space shared?

6. *Accumulating*. What is saved or hoarded instead of shared? How is it done, how are resources counted, what are they weighed or measured against? What is the value of a dollar? How are points accumulated in a game? How do you add or multiply large sums? How do you know you're not being cheated?

7. *Exchanging*. How are things traded? How is value set? How many dolls are worth one bike? How are trading cards accumulated or exchanged? How do you weigh goods, balance them against one another? What systems of weights and measures are used throughout the world? Is it possible to get along without money as a medium of exchange? Is there anything that is so valuable that you would never exchange it?

8. *Gambling*. How is fate dealt with? How are odds set? Why gamble? What is there to lose or gain? What gambling strategies do people use? Do you know anybody who has won? How much? What are the chances of winning? What do people feel about winners and losers? How are the winnings

used? How do you play the numbers, the craps, the horses, cards, roulette, Monopoly?

9. *Making, building and fixing.* How do people in the community make clothes, build houses, fix plumbing, paint, plaster, rewire an apartment? What can you build yourself and what do you need help with from outside the community? Are you sure? How do you fix a car, rebuild one, scrounge around and build one yourself? How have other peoples solved the problem of building their own communities? How can other people's solutions be rethought in your own circumstances? How can you make your own basketball court? Chess or checker game? Wari board? Create your own game? How can you gather together the skills already available in the community, record them and teach them to others?

10. *Dealing with machines.* What machines do you come into contact with every day? How can they be understood? What is the power of people who command the knowledge of how to build and fix machines? How can that knowledge/ power be acquired? What games or toys exist that help understand principles involved in broadcasting, television, understanding and using computers? How does one get hold of resources? Is technology beyond the understanding of most people or is knowledge withheld deliberately to keep people with little power awed?

11. *Curiosity about numbers and relationships as abstractions.* What puzzles do people set for each other or try to solve? How does one go about exploring thoroughly unfamiliar material, trying to solve traditional puzzles or paradoxes?

This last category leads back to the teacher who, by respecting and knowing a community, can offer it unfamiliar material and have the gift accepted rather than suspected — accepted because it helps people to better understand the world and gain control over their lives. The teacher is not a judge of his or her students but rather a worker whose role is to serve their needs and broaden their options. The community and the classroom must help each other.

Recommended reading

There are a number of books that I've found valuable in teaching myself about games and in coming up with ideas that can be translated into classroom activities. These books range from practical handbooks to technical mathematical treatises. Here is a list of these resources, some of which you might find useful:

The World Book of Children's Games by Arnold Arnold (World, 1972, $9.95; Fawcett M1806, 95 cents in paper) is a copious anthology of games collected throughout the world. The eighty pages on strategic games, as well as the sections on outdoor games and word games, can be used in almost any school setting as well as at home and on the streets. This is the most practical and least expensive games book I have found.
Hoyle's Rules of Games edited by Albert H Morehead and Geoffrey Mott-Smith (Signet, 1971, 45p) is cheap and useful, especially as a source of rules for the familiar Western games. The book's emphasis is on card games, though dominoes, checkers and chess receive some attention.

There are two other books that are useful sources of card games:
Teach Yourself Card Games for One by George F. Hervey (Hodder and Stoughton, 1965, 60p).
Teach Yourself Card Games for Two by Kenneth Konstam (Hodder and Stoughton, 1965, 40p).

Perhaps the biggest game freak and one of the most ingenious game inventors in the United States is Sid Sackson. His book,
A Gamut of Games (English edition by Nelson, £2.25) is full of games he has invented as well as a whole range of other recently developed games and traditional game ideas. Sackson's descriptions of games and the strategies they entail are as clear as any I've seen. There is a useful appendix in which Sackson lists and describes commercially available games.
The Master Book of Mathematical Puzzles and Recreations by Fred Schuh (Dover, 1969, $3.00) is a good source of

mathematical games such as NIM and the three pawn game as well as a number of probability games. The book goes into a mathematical analysis of some simple games in detail and is worth reading by anyone who wants to pursue the question of teaching mathematical concepts through games.

There are a series of books that provide introductions to games that are not commonly known in the Western world. Perhaps the most comprehensive is:

A History of Board Games Other Than Chess by H J R Murray (Oxford) is particularly good at analysing some of the strategic complexities of Mancala or Wari.

Other good sources are:

Board and Table Games From Many Civilizations, Vols I and II by R C Bell (Oxford, 1969, Vol I in paperback 75p, Vol II in hardback £2.75).

Games Ancient and Oriental and How to Play Them by Edward Falkener (Dover, paperback £1.68). This is a reprint of an 1892 edition which is particularly valuable because it gives sample games (in notations created by the author) of all the games described. It is easier to get the feel of games this way than by merely reading about their rules.

Puzzles and Games, distributed through the *Punch* group. I feel no games enthusiast can do without a subscription to this magazine. It is full of new games ideas, suggestions to experts and fascinating historical information. So far as I have been able to discover there is nothing comparable to it anywhere in the world.

With respect to specific games, the best introduction to Go I have seen is:

Go for Beginners by Iwamoto Kaoru (Ishi Press, Box 1021, Berkeley, California 94701, $2.50).

Steppingstones to Go by Shigemi Kishikawa (Tuttle, 1965, £1.70).

There is an interesting pamphlet on Mancala (Wari) which was originally printed by the Smithsonian Institute and has been reprinted by Products of the Behavioral Sciences. The illustrations are good, the text an interesting example of colonialist anthropology:

Mancala: The National Game of Africa (available from Products of the Behavioural Sciences, Palo Alto, California).

The literature of chess is vast and any selection has to be highly personal. For people studying games there is an interesting book on the origins of chess-type games and the different forms they have taken throughout the world. There is probably a year's curriculum to be found in the material in this book:
Chess Variations: Ancient, Regional and Modern by John Gollon (Tuttle, 1968, £1.75).

For beginners:
Chess Self-Teacher by Al Horowitz (Barnes and Noble, 1961, 72½p).
First Book of Chess, Al Horowitz and Fred Reinfeld (1953; English edition Faber, 60p).

A good book for those beginning to get interested in chess or for those who want practice in reading chess notations is:
Great Short Games of the Chess Masters by Fred Reinfeld (Collier-Macmillan, 1961, 65p).
My System by Aron Nimzovich (Bell, 1929, £2.75), introduces a very personal approach to chess strategy emphasizing territorial play rather than simple capture of the opponent's pieces.

Two sources for the historical study of games provide models that can be imitated in one's own community. The first one below is also a pleasure to read:
Children's Games in Street and Playground by Iona and Peter Opie (Oxford, 1969, £4.00).
The Study of Games by Elliott Avedon and Brian Sutton-Smith (Wiley, 1971, $10.25).

There are a number of books that deal with the mathematical theory of strategy games. This is an extremely complex matter and the best primer I've seen is:
The Compleat Strategyst: being a primer on the theory of games of strategy, J D Williams (McGraw-Hill, 1965, £7.55).

The classic work on game theory which I must mention, even though I have never made my way through it, may help to explain how the simple study of games has led to some of the technological horrors of simulation theory used in Vietnam and other parts of the world:
Theory of Games and Economic Behavior, John von Neumann and Oskar Morgenstern (Princeton University Press, 1953, £8.90; published in paper by Wiley, £6.00).

Finally, two books for teachers who want to broaden their children's approach to numbers and problem-solving:
The Book of Numbers by Herbert and Mafalda Spencer (Latimer, £2.75) is a book of photographs of numbers in everyday life, and has a 'join-in' section at the end with work projects and number games etc for 7-11-year-olds.
Children Solve Problems by Edward de Bono (Penguin, 1972, 80p) shows the ways in which groups of children solved the problems encountered in nine 'tasks'.

Appendix

(Continued analysis of three-pawn game from page 113.)

There is only one opening move left to analyse, W2 → b2:

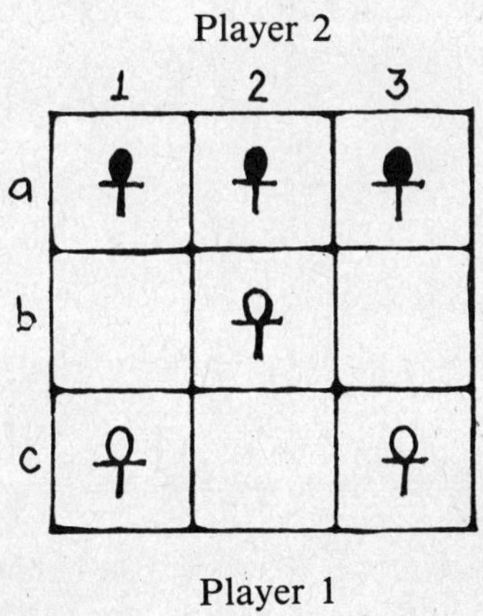

The white piece at b2 has to be taken by BL1 or BL3.* If not, after black's move (say BL3 → b3) white would be able to capture and win (W2 → a1 [X] !). Let's suppose BL1 → b2

*If BL3 — b2 (X), the analysis is identical to the one given here because of the symmetry of the board, which was discussed previously. Try to reproduce the above argument with this move to verify this.

(X). Then the board looks like this:

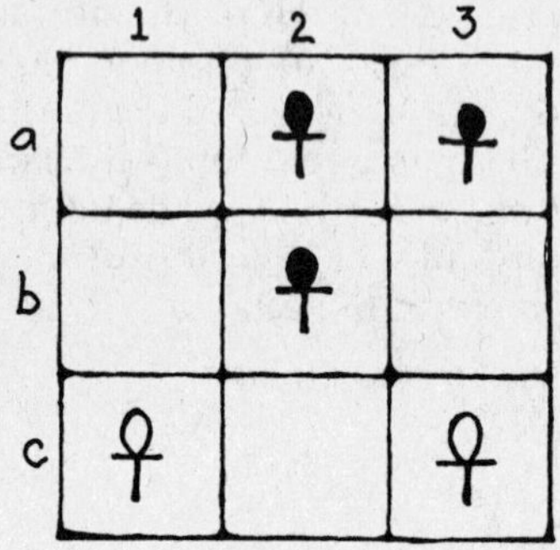

At this point there are a number of possible moves for white to make. Let's consider first the moves of the piece at c3. This piece (W3) can move forward to b3. However, black would reply by BL1 → c2 ! and win. The move left for W3 is to take the black piece at b2, leaving the board:

move:
W3 → b2 (X)

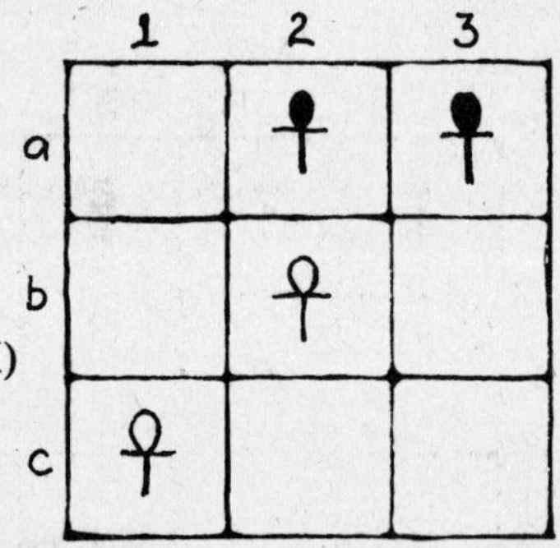

Black can then respond BL3 → b3 and after white's W1 → b1 (the only move possible for white at this point) move BL3 → c3 and win.

So far there is no way that black (the second person to play) can lose if he or she plays intelligently. Now let's consider a final possibility.

When the board is at this position:

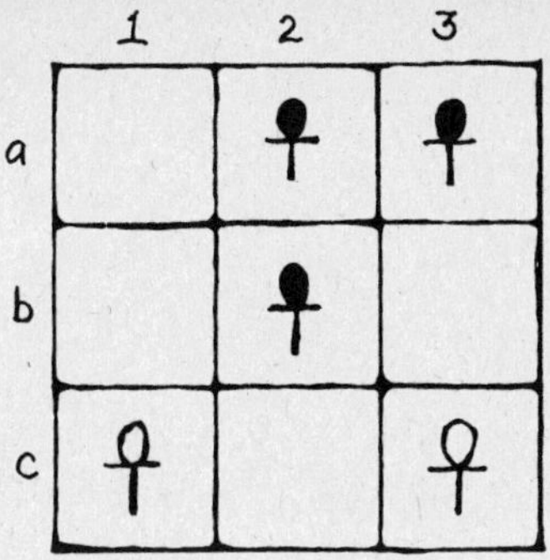

Player 1 can move the piece at c1 instead of the piece at c3 (whose moves were just analysed). The piece at c1 (designated as W1) can move forward to b1. In that case black moves BL1 (on square b2) → c2 ! and wins.

W1 has one other move: W1 → b2 (X). However, then the move BL3 → b3 is also a win since white cannot move:

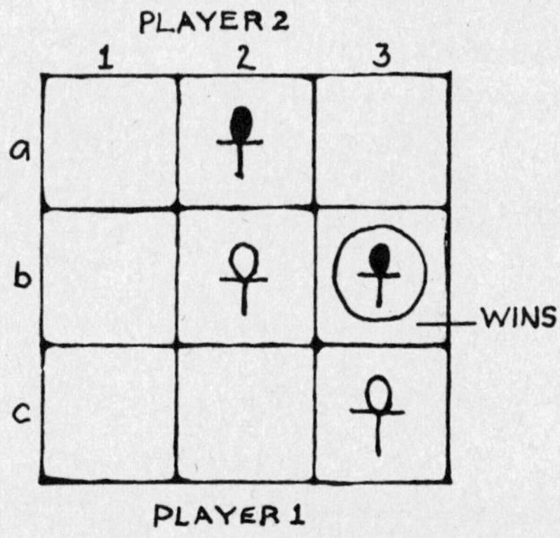

The conclusion of the analysis is that no matter how well the first player plays this game, if the second player makes the best moves available the second player will always win.

HERBERT R. KOHL is the author of *The Open Classroom, 36 Children, Reading, How To, The Age of Complexity* and the pamphlet *Teaching the 'Unteachable'*. He received his bachelor's degree in philosophy from Harvard and a master's degree in special education from Teachers College, Columbia, and was a Henry Fellow at Oxford in philosophy. He has been a teacher for many years in Harlem and Berkeley, where he was the director of Other Ways, a much admired alternative public school, and now teaches kindergarten and first grade in Berkeley. He has been the director of Teachers and Writers Collaborative, a group that is revising the curriculum in elementary and secondary schools, and a columnist for Grade Teachers Magazine, and is a contributor to *The New York Review of Books.*

Notes